Geology and Landscape
of Michigan's
Pictured Rocks
National Lakeshore
and Vicinity

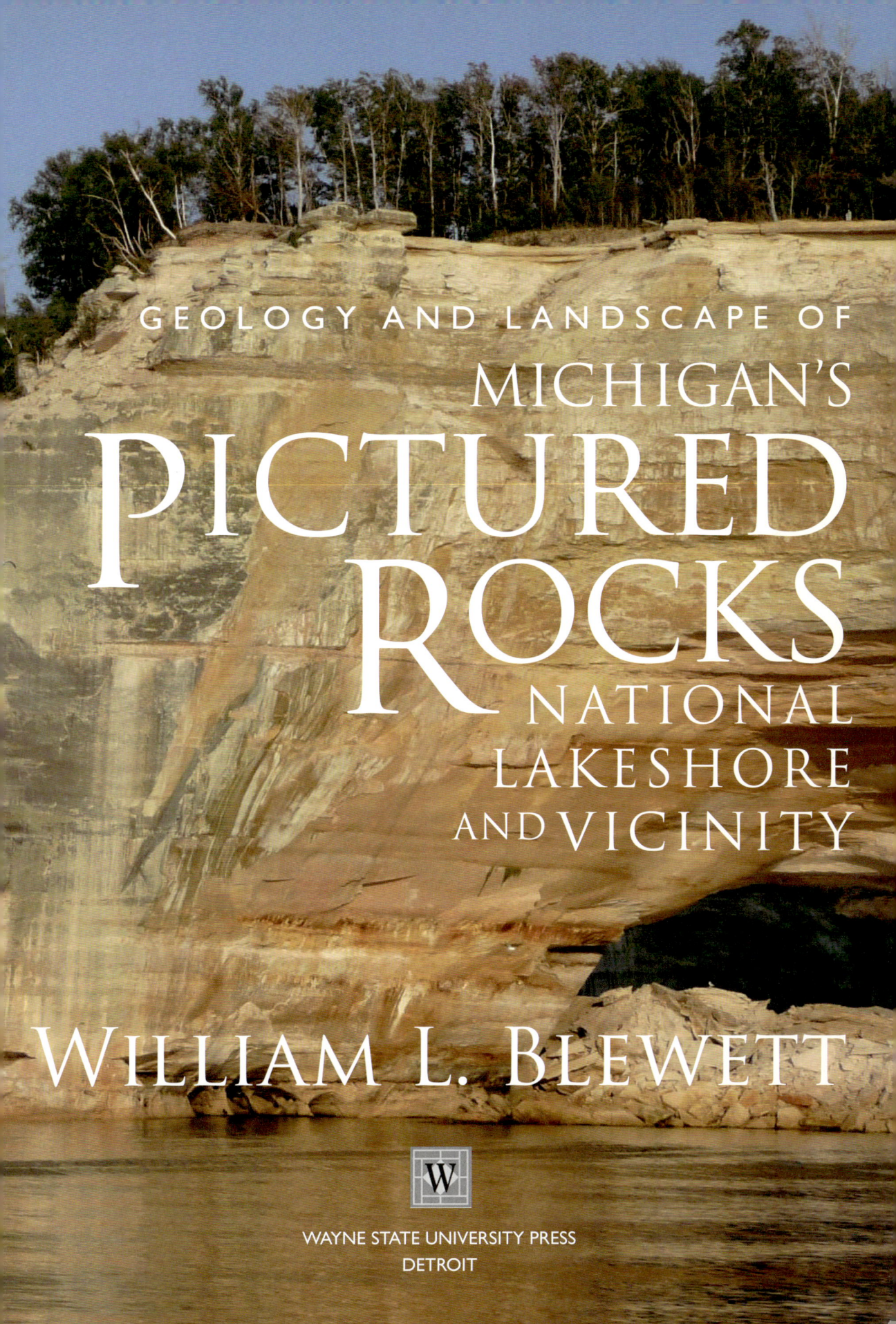

GEOLOGY AND LANDSCAPE OF

MICHIGAN'S
PICTURED
ROCKS
NATIONAL
LAKESHORE
AND VICINITY

WILLIAM L. BLEWETT

WAYNE STATE UNIVERSITY PRESS
DETROIT

16 15 14 13 12   5 4 3 2 1

Library of Congress Cataloging-in-Publication Data

Blewett, William L., 1959–
  Geology and landscape of Michigan's Pictured Rocks National Lakeshore and vicinity / William L. Blewett.
     p. cm. — (Great lakes books)
  Includes bibliographical references and index.
  ISBN 978-0-8143-3441-6 (pbk. : alk. paper)
  1. Geology, Structural—Michigan—Pictured Rocks National Lakeshore. 2. Glaciology—Michigan—Pictured Rocks National Lakeshore. 3. Paleogeography—Pleistocene. 4. Paleogeography—Holocene. 5. Geology, Stratigraphic—Pleistocene—Michigan—Pictured Rocks National Lakeshore 6. Geology, Stratigraphic—Holocene—Michigan—Pictured Rocks National Lakeshore. 7. Pictured Rocks National Lakeshore (Mich.) I. Title.
  QE627.5.M5B54 2012
  557.74'932—dc23
                    2011024063

∞

Publication of this book was made possible through the generosity of the Ford R. Bryan Publication Fund.

Designed and typeset by Brad Norr Design
Composed in Minion, Trajan, and Gill Sans

CONTENTS

Preface

This book was a long time coming. It began in 1994 with the writing of a Pictured Rocks Resource Report on the park's glacial geology, followed by a more ambitious report on ancient shorelines written as a "geoscientist-in-the-park" during my sabbatical leave in 2004. Near the end of my stay, Gregg Bruff (the park's chief of Heritage Education), suggested using the two reports as the basis for a book on park geology. I accepted the challenge with some trepidation, knowing full well that my academic responsibilities would slow the project considerably. This turned out to be something of an understatement, but over the next five summers (between committee meetings), I was able to finally finish the manuscript.

My charge is made somewhat easier by the fact that an extensive literature exists on the park. Indeed, the geologic history of Pictured Rocks National Lakeshore is generally well understood by scientists, but much of the relevant literature is unpublished, outmoded, or scattered among a variety of journals, resource reports, and government documents. Accordingly, this book is designed to synthesize all published and available unpublished information on the park's geologic history into one source, in what is hoped will be a lively, lucid, and richly illustrated narrative. It is targeted for the interested public but assumes at least a rudimentary understanding of basic geologic principles as might be taught in an introductory college geology or physical geography class. As such, the book should also be useful to geologists, physical geographers, and those working in closely related fields such as archaeology, biology, ecology, and environmental science. A detailed mileage-referenced road log is provided in the appendix to guide readers to the best and most accessible field sites, and, for the more adventurous, a day hike keyed to the geology is included. These features are augmented by a comprehensive references list located at the end of the book.

The manuscript's organization is chronological. After a short introductory chapter establishing the park's geologic setting, we begin with the oldest rocks in the park and work our way to the present. The geology logically divides into bedrock and drift, and I've tried to present a balanced approach that gives emphasis to both. Some hard rock geologists may be disappointed with the attention given to the surficial deposits and landforms, but the park's landscape (including Lake Superior) is essentially glacial and postglacial in origin, and Pleistocene and Holocene events are an important part of the story. My effort to broaden the appeal of the manuscript also requires that some basic geologic principles be explained at a length appropriate to their complexity. To the geologically proficient, these passages may seem like annoying remediation, and I ask for their patience and understanding.

Finally, I am acutely aware of the dangers inherent in paraphrasing and simplifying the work of other scientists, many of whom are my friends and coauthors. I am also sensitive to my colleagues who might bristle at the thought of someone profiting from descriptions of their scientific work. To this charge I can only say that the book is written at the behest of the National Park Service to elucidate and inform, that books of this sort rarely make a profit, and that it is my ultimate intention to return any remunerations to Pictured Rocks through the Lake Superior Foundation. For any misinterpretations or factual errors, of course, I take full responsibility.

ACKNOWLEDGMENTS

Immeasurable thanks are due to Gregg Bruff, chief of Heritage Education at Pictured Rocks National Lakeshore, who provided the initial inspiration for this book and shared access to innumerable National Park Service photographs and related documents. Likewise, Dr. Walt Loope of the U.S. Geological Survey was extremely generous in sharing materials on the Holocene history of the park, including the many fine photographs found in chapter 5. He also reviewed an earlier version of the entire manuscript and saved me from many embarrassing blunders. Dr. John Anderton of the Department of Geography at Northern Michigan University read large portions of the manuscript, including all of chapter 5, and corrected me on a number of key points related to coastal landforms, park archaeology, and the Sawmill Culture. Dr. Sean Cornell, my geoscientist colleague in the Geography–Earth Science Department at Shippensburg University of Pennsylvania, served as my hard rock geology expert throughout the writing of the book and patiently answered my annoying questions related to the Precambrian and Paleozoic geology of the midcontinent. He also reviewed an earlier version of the entire manuscript. Thanks also to the two anonymous outside reviewers who performed extremely thorough evaluations of the entire manuscript, and to two anonymous reviewers from the Great Lakes Books Series Editorial Board. Their thoughtful comments improved the book significantly.

The National Park Service, through its "Geoscientist-in-the-Parks" program, provided lodging and logistical support for four months at Sand Point during the fall of 2004. They also offered me free reign of the park library, where I was able to examine many of the more obscure and hard-to-find reports on park geology, ecology, and natural history. These efforts were augmented by a grant from Eastern National Corporation (the company that runs National Park Service book stores east of the Mississippi River), which supported the writing of a park service Resource Report that is the basis for chapter 3. My academic home, Shippensburg University of Pennsylvania, provided me with a sabbatical leave in fall 2004 as well as a generous professional development grant in 2007 to support field work and the actual writing of the manuscript. Chapter 4 is based on a Resource Report written in 1994 that was supported by several additional grants provided by the Pennsylvania State System of Higher Education and my university.

I am also grateful for the many kindnesses, both large and small, offered by so many people during the preparation of this book. Kathleen Weessies of the Michigan State University Library and Julie King of the Library of Michigan's Rare Book Room were instrumental in securing a digital version of figure 2.21 from Foster and Whitney's 1851 report. Shippensburg's Geography Department secretary Judy Ferrell performed innumerable administrative tasks with her trademark spirit of helpfulness and good cheer, while Shippensburg University graduate student Kaja Spaseff assisted with early versions of some of the graphics. Diane Kalathas of the Shippensburg University library helped me locate a number of hard-to-find references. My departmental colleague Dr. Scott Drzyzga helped me with several technical issues related to the graphics. Thanks are also due to Drs. Randy Schaetzl and Dave Lusch of Michigan State University for encouragement and sharing of materials, to Lou Waldock and Tony Williams for permission to use their photographs in figures 5.9 and 5.17 respectively, and to Brenda St. Martin, administrative assistant at Pictured Rocks National Lakeshore, who let me abuse the copier at park headquarters and never turned me in.

A special thanks to all of the crew at Wayne State University Press, whose professionalism, skill, and good sense have made this book better than it was ever destined to be. Of special note are editor-in-chief Kathy Wildfong, who shepherded the book through its initial review; assistant editorial manager Carrie Downes Teefey; assistant design and production manager Maya Rhodes; and the sales and marketing team of Emily Nowak, Sarah Murphy, and Jamie Jones. Freelance copyeditor Dawn Hall was an extremely efficient taskmaster and caught more mistakes than I ever thought possible in the Age of Spellchecker. Freelancer Jane Henderson created an excellent index.

Most of all, I am indebted to the small group of scientists whose work forms the basis for our present understanding of Pictured Rocks National Lakeshore. These include Bill Hamblin, Charles Haddox, Bob Dott, John Hughes (my former professor at Northern Michigan University), Walt Loope, Tim Fisher, John Anderton, Chris Drexler, and the late Pat Farrell (another of my former NMU professors). I also wish to thank geologist Bob Rosé, whose interpretative materials on the Munising and Au Train Formations completed during his own Geoscientist-in-the-Park experience at Pictured Rocks were extremely helpful to me in clarifying outcrop patterns in the Munising Formation, and form the basis for several of the book's illustrations.

Finally, I must thank my wife, Gretchen. Having lived through the completion of a Ph.D. dissertation many years ago, she can now feel confident that she has paid her debt to society twice over and can look forward to a leisurely retirement with time off for good behavior.

1

INTRODUCTION AND REGIONAL SETTING

INTRODUCTION

Pictured Rocks National Lakeshore preserves one of the most exquisite freshwater coastal landscapes in North America. Here, nearly five hundred miles from the nearest ocean, lies a lonely, 40-mile swath of jagged coastline to rival any New England travel brochure (Fig. 1.1). The Pictured Rocks—soaring sandstone fortresses awash with pink, green, and brown natural pigments—anchor the park's western end, giving way to a seemingly endless sandy strand arcing eastward to Grand Sable Dunes. The dunes perch precariously atop a steep, nearly three-hundred-foot-high embankment of glacial debris called the Grand Sable Banks, offering a surreal desert vision set against the endless horizon of the world's largest freshwater sea.

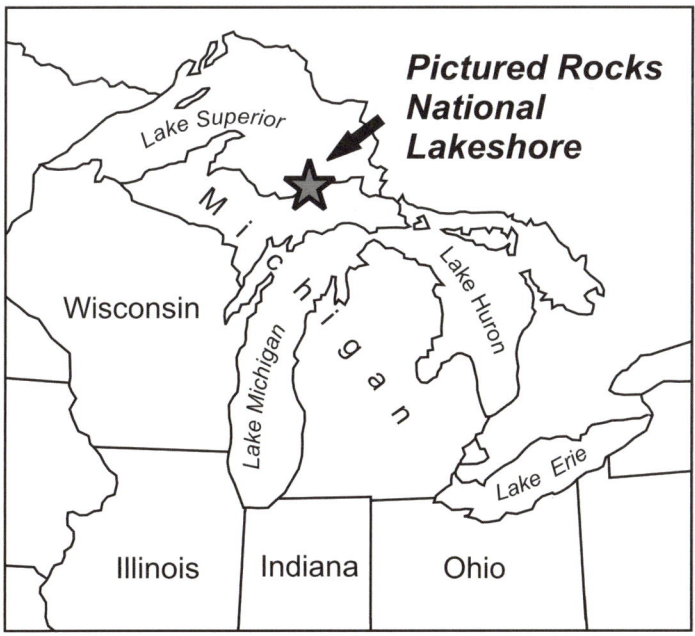

FIGURE 1.1. Locator map for Pictured Rocks National Lakeshore.

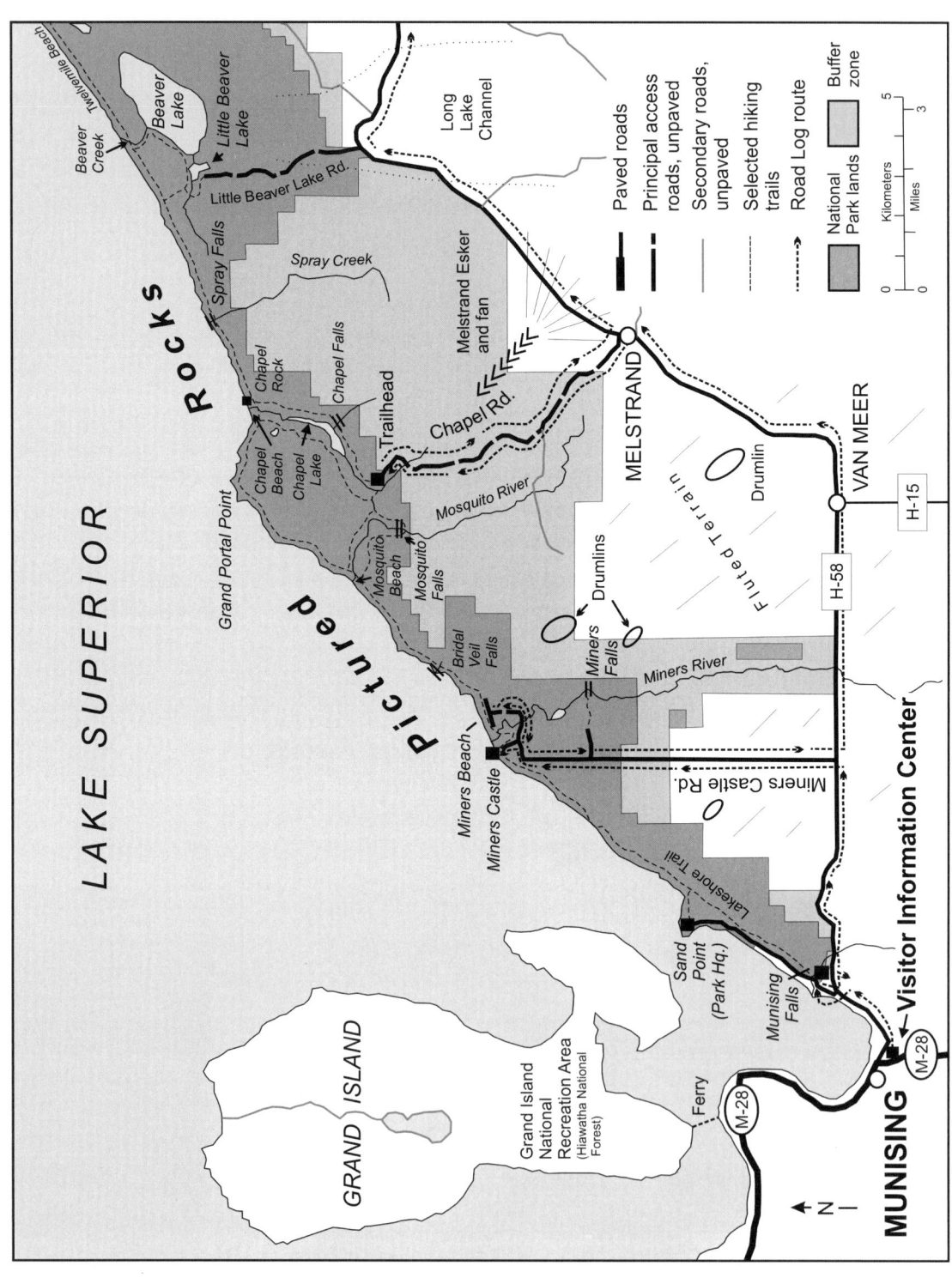

FIGURE 1.2a. Map of Pictured Rocks National Lakeshore and vicinity (west half).

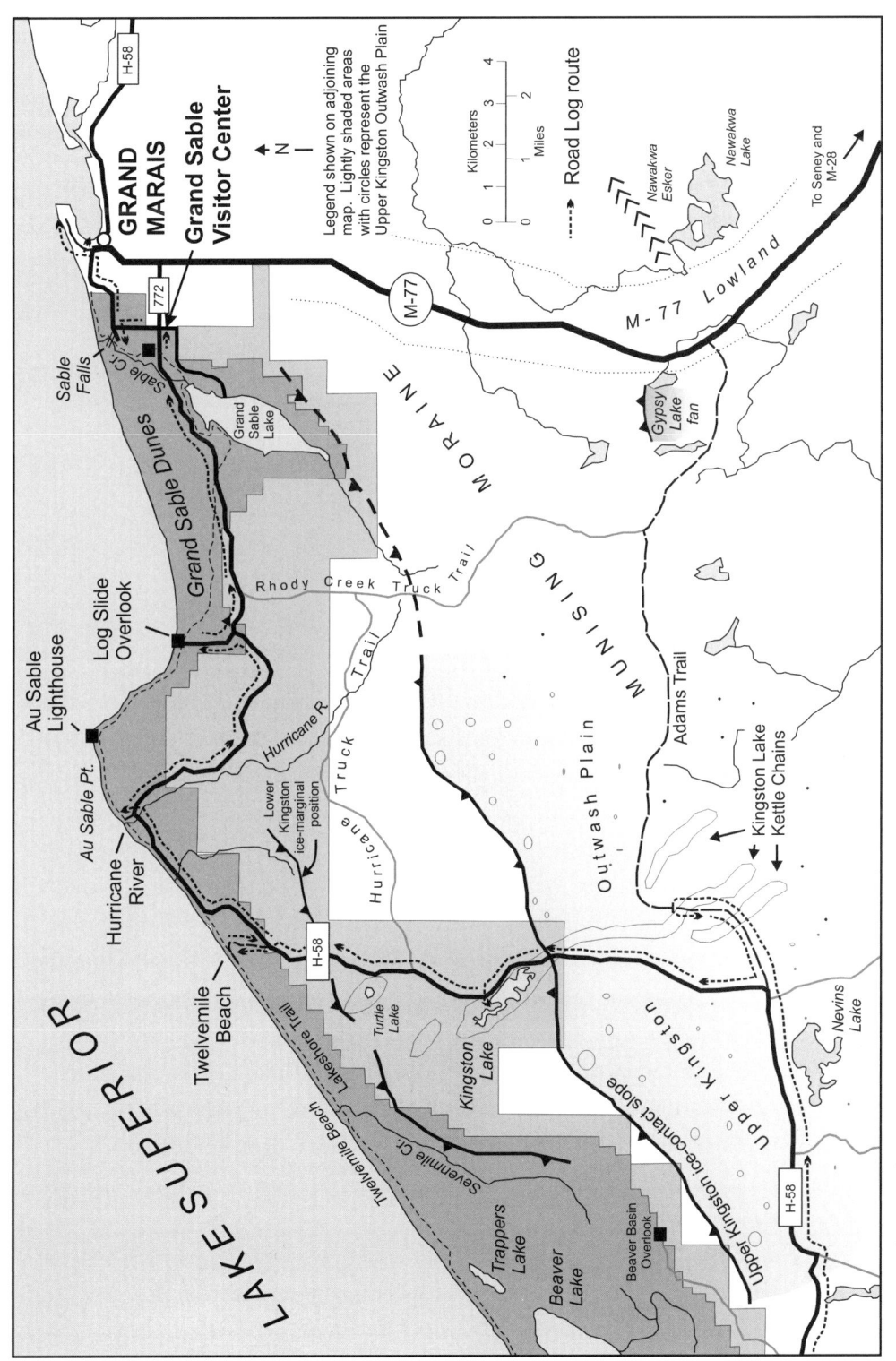

FIGURE 1.2b. Map of Pictured Rocks National Lakeshore and vicinity (east half).

The park is immense by midwestern standards and encompasses 33,929 acres along a narrow 43-mile strip of mostly wilderness coastline between Munising (pronounced: MEW neh sing) and Grand Marais (ma RAY), Michigan (Figs. 1.2a, b). An additional 39,306-acre buffer zone, where development is strictly regulated, fringes the park's southern boundary. The Pictured Rocks cliffs (Fig. 1.3) and Grand Sable Dunes (Fig. 1.4) are the park's principal attractions and are considered by some to be among the most striking natural features in eastern North America (Karamanski, 1995). Other park attractions include Miners Castle, Munising Falls, Miners Falls and Beach, Chapel Rock and Beach, Twelvemile Beach, the Lakeshore Trail, and Sable Falls.

To the student of geology and landforms, the Pictured Rocks region provides an excellent natural laboratory for exploring the materials, processes, and events involved in shaping this most sublime of midwestern landscapes. This book is designed to provide a general overview of our current understanding of this extraordinary terrain and to guide the interested visitor to the best and most accessible field sites.

ACCESS

The Pictured Rock cliffs, while visible from land, are best viewed by either private or commercial boat or by kayak. First-time visitors are encouraged to use the excellent commercial boat tours available at the dock in downtown Munising. The tours are available from mid-May to early October and are professional, informative, and safe. Depending upon weather conditions, they also offer up-close views of the rocks to satisfy even the most demanding sedimentary petrologist. Because Lake Superior water temperatures rarely exceed 50 degrees Fahrenheit, a wool sweater or jacket is highly recommended for the two-hour boat ride. Private boaters should begin planning their tours at the National Park Visitor Information Centers in Munising or Grand Marais, as some restrictions apply. The centers also contain up-to-date weather information. Inexperienced boaters or kayakers should not attempt to navigate the open waters of Lake Superior. Weather conditions at this northern latitude are notoriously fickle, and survival times in Superior waters are very short, even in summer. No emergency landing zones exist for a distance of nearly 8 miles along the Pictured Rocks cliffs, and small boats can be easily swept against the cliffs in a sudden squall.

Land access is on foot via the Lakeshore Trail or by Alger County Road H-58, which traverses the park from west to east at varying distances from the coast (Figs. 1.2a, b). Roads emanating northward from H-58 intersect the shoreline at a few spots (Miners Castle, Miners Beach, Twelvemile Beach, Hurricane River, Log Slide), but

FIGURE 1.3. The Pictured Rocks cliffs, looking southwest at sunset.

automobile access to shore points is limited. Other roads bring visitors to trailheads within a few miles or less of the coast and are excellent ways to reach shoreline outcrops and beaches (Mosquito Beach, Chapel Beach, Twelvemile Beach via Little Beaver Lake, Sable Falls and the mouth of Sable Creek). As this book goes to press, H-58 is now paved from Munising to Grand Marais (Figs. 1.2a, b). Gone are the adventurous days of washboard gravel roads and skidding logging trucks. To many purists (including this one), the pavement has irrevocably changed the nature of the Pictured Rocks experience. For the student of geology, however, this means more comfortable automobile access to the park interior and selected shore points, but the

FIGURE 1.4. Grand Sable Banks and Dunes, looking east from the Log Slide overlook.

number of geologic sites accessible by automobile remains limited. The best way to see the rocks continues to be by boat (Fig. 1.5).

For the more adventurous, the 40-mile-long Lakeshore Trail hugs the coast from Munising to Grand Marais and is the best way to experience the park geology from land. Many of the sites described in this book are accessible from the Lakeshore Trail. The appendix provides a road and trail log tied to the geology. As with any northern wilderness, black flies and mosquitoes can easily test the sanity of even the most well adjusted geologist, especially in late June. Quality insect repellant and/or head netting can make your visit to that special outcrop much more enjoyable.

Nearby Grand Island is part of the Hiawatha National Forest and is administered as a recreation area. Its cliffs contain excellent exposures of the Jacobsville and Munising Formations that are best viewed by private boat or kayak. Public access to the island is provided by a small pontoon ferry service that operates from late May to early October from a dock at Grand Island Landing, about 3 miles west of Munising off M-28 (look for the sign). Travel on the island is by bicycle, foot, or tour bus.

HISTORY AND MISSION OF THE PARK

Pictured Rocks National Lakeshore was established on October 15, 1966 as the first U.S. National Lakeshore, followed in rapid succession by Indiana Dunes

FIGURE 1.5. Spray Falls, located near the eastern end of the Pictured Rocks cliffs, is best viewed by boat.

(1966), Apostle Islands (1970), and Sleeping Bear Dunes (1970). Local sportsmen's groups in the eastern Upper Peninsula had been lobbying the State of Michigan since 1923 to initiate a park in the area under a number of different guises and proposed configurations. In Munising, organized interest was not so much in the Pictured Rocks cliffs, but in the so-called Beaver Basin east of the cliffs, which was an extraordinarily rich and remote ecosystem supporting a large deer population. Boosters in Grand Marais lobbied for a park centered on the Grand Sable Banks and Dunes as a way to augment the emerging Northwoods tourist industry. For the most part, these initiatives were driven by hunting, fishing, and economic interests rather than concerns for preserving natural ecosystems (Karamanski, 1995).

The federal government became interested in fostering recreational opportunities in the Great Lakes states in the late 1950s and commissioned a shoreline survey to identify potential sites. Pictured Rocks was recognized as an area of exceptional potential, along with Sleeping Bear Dunes, the Huron Mountains, and Indiana Dunes. Accordingly, the park service proposed a 100,000-acre park stretching from Munising in the west to Grand Marais in the east. Most citizens in Grand Marais and Munising reacted favorably to the proposal, but lumbering interests, specifically the Cleveland-Cliffs Corporation who owned much of the land on which the park would be located (and would be an essential partner in making the park a reality), objected strongly to the proposal. They were concerned that such a large park would remove too much timberland from harvest. Through complicated maneuvering over several years, a compromise was eventually worked out whereby the park service would take complete control of a smaller, narrower strip of land containing the principal natural attractions immediately along the lakeshore, with a mixed-use buffer zone along the park's southern boundary in which timber harvesting and other uses would be permitted under strict regulation (Karamanski, 1995).

The park's initial mission was born of the Great Society and the spirit of development and use. The park was sold to locals as a means of economic development rather than as preservation. Park planners proposed intensive recreational shoreline development capable of supporting 1 million visitors annually, including a shoreline drive along the crest of the cliffs. Later, the shoreline drive was to prove one of the most controversial aspects of the proposed plan, and it continues to haunt the Lakeshore in somewhat reduced form today.

The budgetary realities of the Vietnam War and growth of the preservationist movement caused these initial plans to be shelved and eventually reevaluated. Budgetary constraints kept development within the park to a minimum until well into the 1970s. Major environmental laws such as the Clean Air Act, Clean Water Act, Endangered Species Act, and the National Environmental Protection Act also constrained options for park development. This perceived "foot-dragging" by the National Park Service led to growing animosity between the park service and the local citizenry, many of whom felt that the slow progress was in fact a cleverly disguised plot by the park service to pursue a preservationist agenda. Local businessmen bemoaned the lack of improved campgrounds and access to lakeshore points, complaining of backpackers who "come to town with a five dollar bill and a pair of underwear and never change either one." In 1977 an attempt was made to rectify the situation by forming a task force consisting of local citizens, business interests, and park personnel. This group made a number of recommendations to the park service that were then incorporated into a revised general management plan in 1981. The proposed shoreline road was downgraded to one that followed existing roads, most notably County Road H-58, with paved spurs to shoreline points of particular interest. Notably, surveys of park users indicated that the vast majority (97 percent) favored minimal development and opposed the shoreline road.

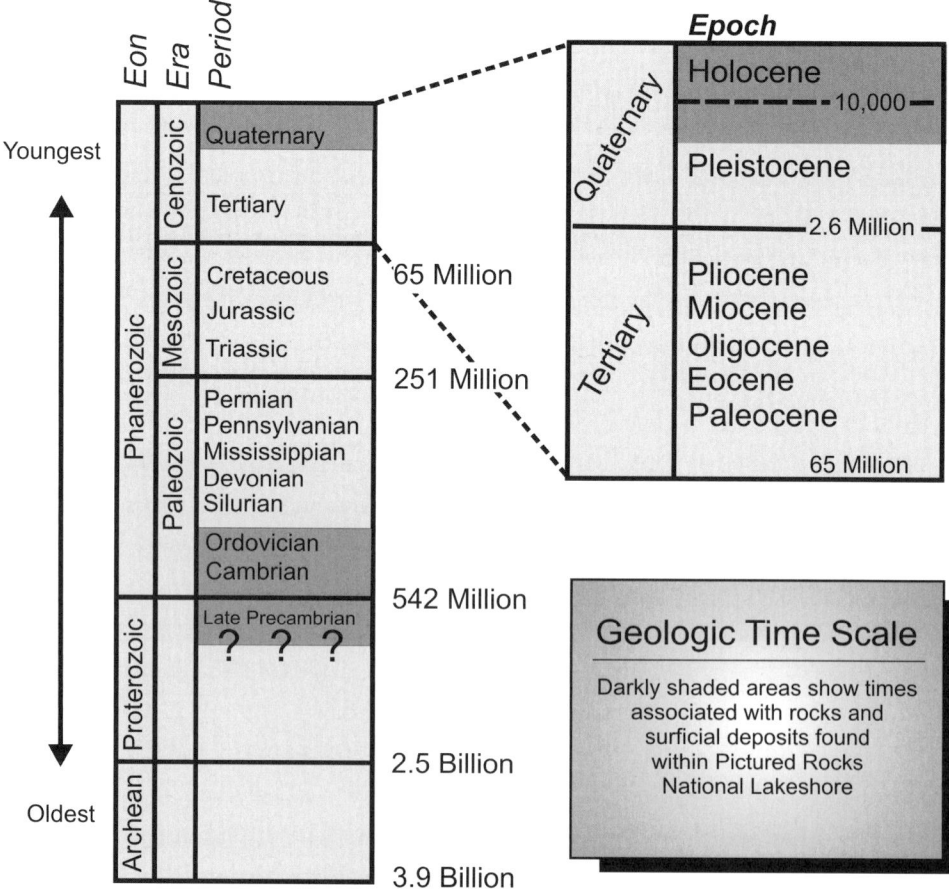

FIGURE 1.6. The Geologic Time Scale (after Walker and Geissman, 2009).

This disconnect between park users and local economic interests was an ongoing management challenge within Pictured Rocks during the 1970s and 1980s. Local attitudes have changed, however. According to a former park manager and current Munising resident, the days of fierce local opposition to the national lakeshore have faded, replaced by a growing respect and appreciation for the park.

In 2004 the park service completed a new general management plan that stressed more convenient access to eastern and western ends of the park while preserving central portions of the park in an undisturbed and primitive state. The paving of H-58 during the summer seasons of 2008–10 was a major component of this plan and, while providing improved access, has dramatically changed the nature of the Pictured Rocks experience.

Pictured Rocks National Lakeshore celebrated its fortieth anniversary in October 2006. From its slow beginnings, the park has matured following a series of compromises among business interests, the local populace, park visitors, and the National Park Service. The park has never reached the million annual visitors projected by park planners back in the 1960s, but visitation has increased markedly over the years, reaching 443,370 visits in 2007. With the paving of H-58, one of the most significant goals of the current master plan is now in place. The continuing debate regarding preservation versus development will likely continue for the foreseeable future.

GETTING STARTED: THE GEOLOGIC TIME SCALE AND PICTURED ROCKS STRATIGRAPHY

Geologists tell time using the geologic time scale. As you may remember from science class, they divide the time since the Earth began into large chunks called eons, which are further subdivided in descending order of magnitude into eras, periods, and epochs (Fig. 1.6). Few of these time periods are represented by the rocks in Pictured Rocks, in fact, if the rock record at Pictured Rocks were a book, it would be one in which only the first and last chapters were intact, with all the intervening chapters missing. This situation allows us to separate our description of the geology neatly into two parts: the very ancient Precambrian, Cambrian, and Ordovician rocks dating back hundreds of millions of years (chapter 2), and the much younger, unconsolidated (loose or uncemented) Pleistocene and Holocene sediments mantling the bedrock (chapters 3, 4, and 5), most of which are no older than 12,000 years (Fig. 1.6).

Geologists use the term *formation*, the fundamental descriptor in geologic taxonomy, to refer to closely related heterogeneous rock units with a similar history. These are often further subdivided into *members*, which are distinctive rock layers within a formation that are more or less homogeneous throughout.

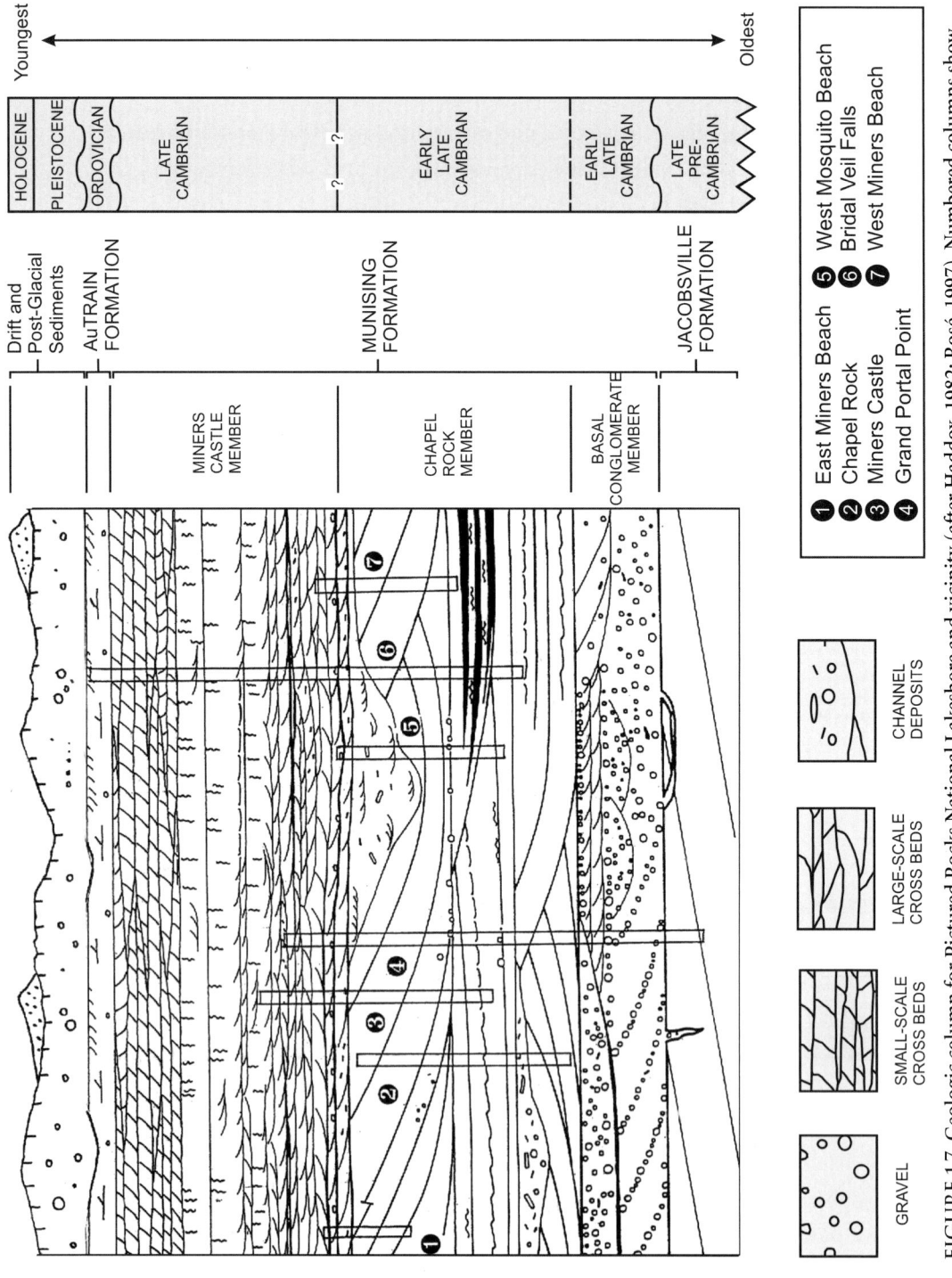

FIGURE 1.7. Geologic column for Pictured Rocks National Lakeshore and vicinity (after Haddox, 1982; Rosé, 1997). Numbered columns show the extent of bedrock exposures at the listed sites.

Using radiometric dating, fossil assemblages, regional correlation, and other means, geologists assign the rocks in a given area to particular time periods, creating a *geologic column* for that region. The various formations are stacked vertically, with the oldest layers on the bottom and the youngest on top, mimicking their typical arrangement in the field.

Figure 1.7 shows the geologic column for Pictured Rocks National Lakeshore and vicinity. The Jacobsville Formation (approximately 1.1 billion to 670 million years old), a red sandstone of Precambrian age, is the oldest rock exposed in the park. The Late Cambrian Munising Formation (480 to 500 million years old) lies unconformably above the Jacobsville and consists of three members: a basal conglomerate, the Chapel Rock sandstone, and the Miners Castle sandstone. The term *unconformable* means that a significant period of erosion, possibly representing millions of years, occurred between deposition of the older sediment and the overlying strata. Capping the Munising Formation is the resistant Au Train Formation of Early Ordovician age. This light brown to white dolomitic sandstone is the most widespread lithology (rock type) exposed at the surface in the park, although much of it is covered by glacial debris. The Pictured Rocks cliffs are developed upon the Munising and Au Train Formations. The formations are best viewed at the shoreline or in scattered locations farther inland where streams or glacial meltwater have removed the thin overburden. Overall, each of the Paleozoic formations dip gently south toward the center of the Michigan structural basin (see below), a large, regionally significant center of marine accumulation

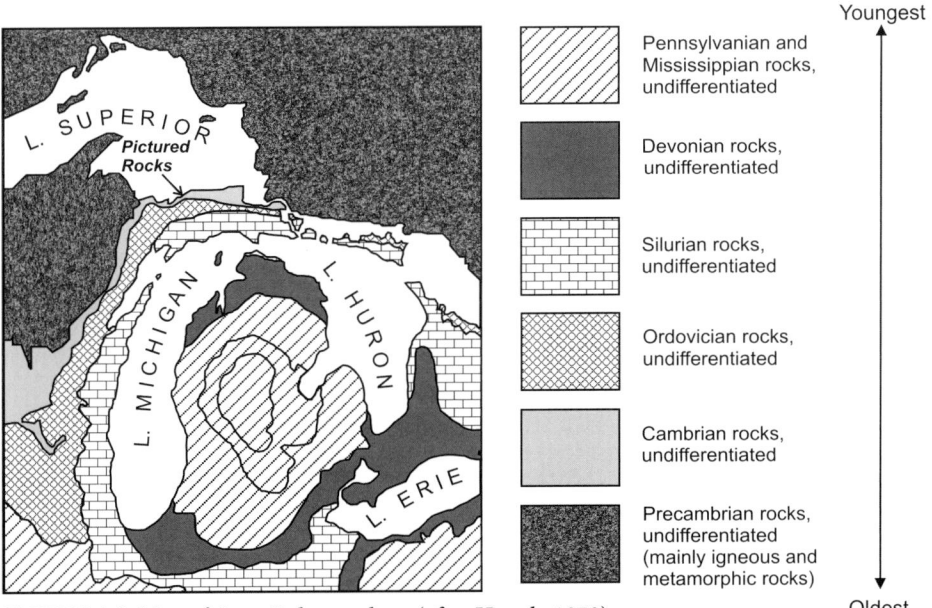

FIGURE 1.8. Map of Great Lakes geology (after Hough, 1958).

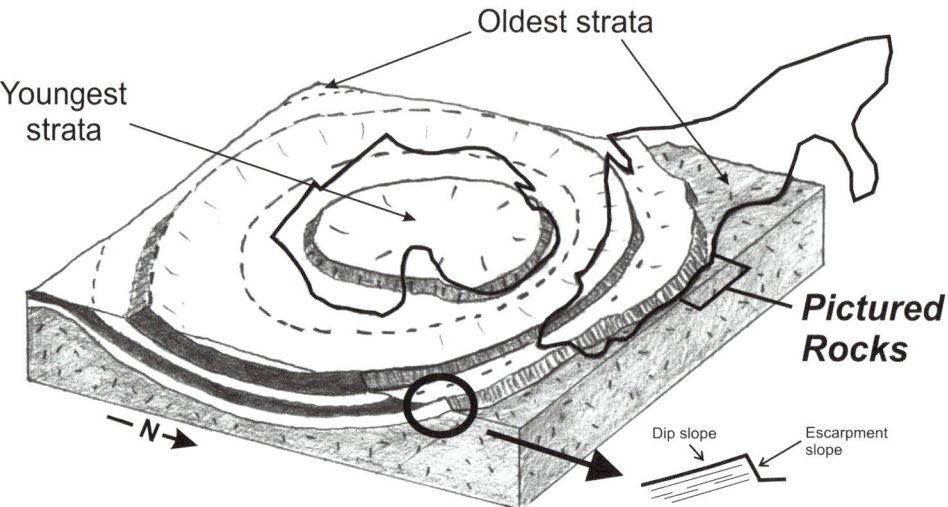

FIGURE 1.9. Idealized diagram showing the general relationship of the Pictured Rocks escarpment to rocks of the Michigan structural basin. Stratigraphic details are highly simplified and not to scale. The first demonstrable connections between the north-central Upper Peninsula and the Michigan Basin date to the Ordovician. Drawing by the author.

associated with a succession of shallow Paleozoic seas (Figs. 1.8, 1.9, 1.10). These units also tip slightly westward, so that the older Jacobsville Formation (on the bottom of the stack) is best exposed in eastern sections of the park.

Unconsolidated Pleistocene and Holocene sediments of varying thickness mantle bedrock throughout the region. *Unconsolidated* means that the materials are loose rather than hard bedrock. Glacial sediment (called *drift*) is assigned to the very latest phases of the Wisconsin glacial episode, the last major ice sheet to affect the Great Lakes area (Drexler and others, 1983; Blewett, 1994). Details of Pleistocene and Holocene sediments and events are reviewed in chapters 3, 4, and 5.

GEOLOGIC SETTING

The Pictured Rocks are part of a regionally extensive, northward-facing escarpment that can be traced nearly two hundred miles, either at the surface or beneath the drift, from Marquette eastward to Sault Ste. Marie. Escarpments are cliff faces that form where sedimentary rocks have been tilted from the horizontal. The tilting forms a ridge that is asymmetrical in profile, with a steep face called the *escarpment slope* on one side, and a gentler slope called the *dip slope* on the other (Fig. 1.9). The direction the escarpment slope faces is determined by the relationship of the rocks to the overall regional structure, in our case, the downwarped strata of the Michigan structural basin. This area, centered on the Lower Peninsula of Michigan, was a depressed area of the crust, which served as the center of accumulation for thousands

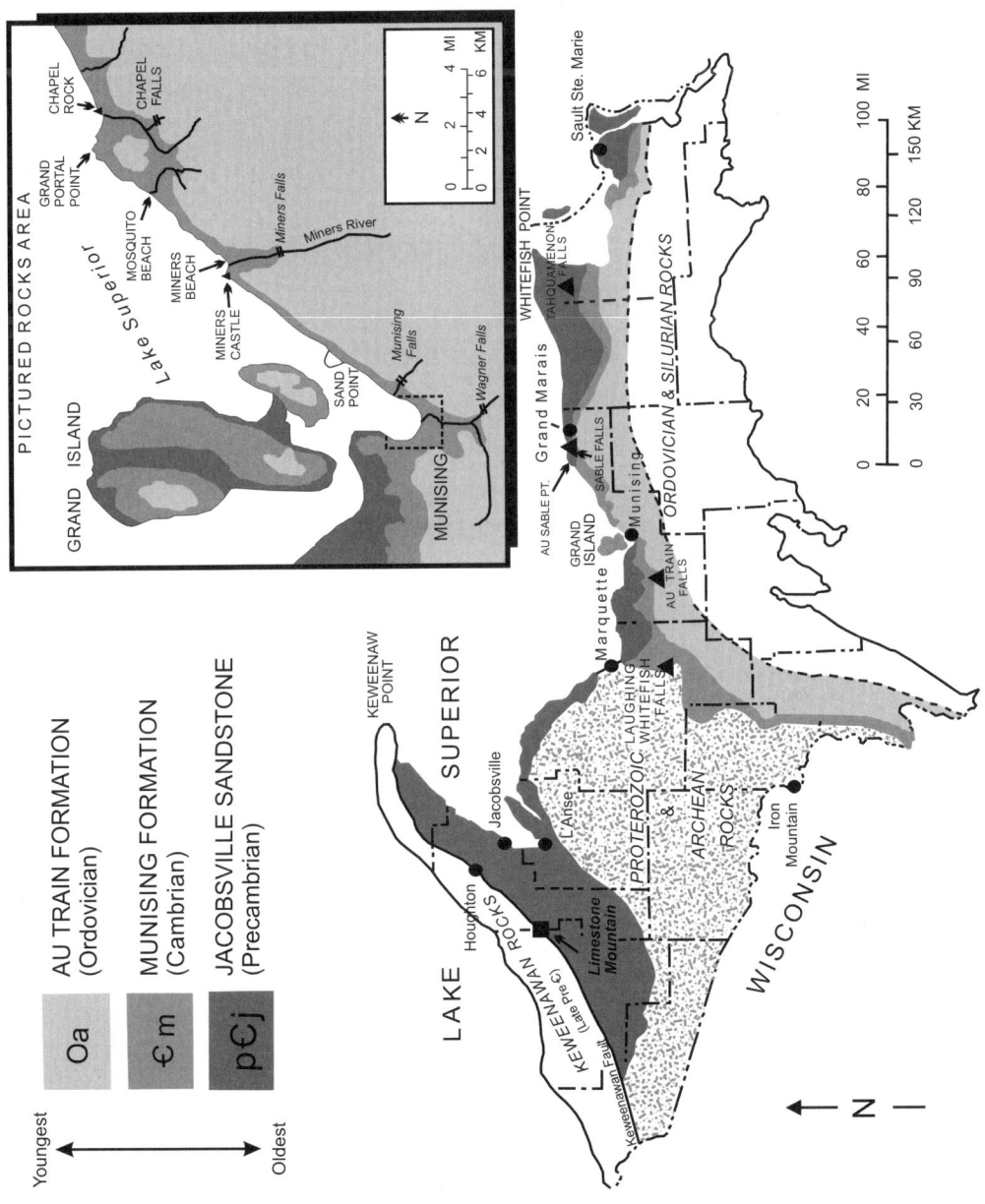

FIGURE 1.10. Map of Upper Peninsula geology (after Haddox and Dott, 1990).

of feet of marine sediments during the Paleozoic era. The pattern these now lithified sediments display can best be imagined as a set of nested bowls, set one inside the other. The bottom-most bowl represents the oldest strata, with the top bowl representing the youngest. If the bowls are viewed from above, a distinctive target pattern results, with the upturned edges of the outer (or oldest) bowl exposed along the outer circle of the target, and with the innermost (or youngest) bowl exposed at the center. This pattern can be clearly discerned in the geologic map of Michigan shown in figure 1.8. Note that the rocks of the Munising (labeled as "Cambrian rocks, undifferentiated") and Au Train Formations ("Ordovician rocks, undifferentiated") are the oldest of the Paleozoic units shown and thus are located at the periphery of the structural basin. Because some of the rock layers in a basin are more resistant than others, however, weathering and erosion will tend to form ridges on the more resistant rocks and valleys or lowlands upon weaker rocks. Notice in figure 1.9 that the ridges formed in this manner create distinct escarpments with the steep side facing outward, away from the center of the basin. Thus, in a structural basin, the oldest formations will be found at the outer margins of the basin and will form outward-facing escarpments if they are resistant. This pattern is directly analogous to the northward-facing Cambrian-Ordovician escarpment of the Pictured Rocks. The reality of the situation is somewhat more complicated, however, since the Munising Formation pinches out on the basement rocks of the central Upper Peninsula and cannot be traced with certainty southward into the Michigan Basin (Haddox and Dott, 1990).

THE PICTURED ROCKS LANDSCAPE

Like all scenery, the Pictured Rocks landscape is the result of a complicated interplay among the lithology (rock type), structure (geometric arrangement of the rock), and the processes (weathering, erosion) acting upon the rock and structure. Generally speaking, weathering and erosion (the driving forces) go to work upon rocks and structure (the resisting forces) to produce landforms. The distinction between the "rocks," and the resulting "scenery" developed upon those rocks, is crucial to understanding the Pictured Rocks landscape, because the two are often confused and of enormously different ages. For example, the rocks of the Pictured Rocks cliffs date to the early Paleozoic and are hundreds of millions of years old, yet the actual cliff developed upon these rocks—the scenery—formed through weathering and erosion within only the last few thousand years. One of the fundamental errors made by most visitors is to assume that because the rocks are hundreds of millions of years old, the cliffs are, too. They are, in fact, confusing the age of the landform with the age of the rock that makes up the landform. The folly of this view can be easily demonstrated when one considers that the entire region emerged from beneath the glaciers in only the last 12,000 years. Thus, no landscape feature within Pictured Rocks National Lakeshore can be older than 12,000 years.

Physical geographers have divided North America into regions of similar topography called *physiographic provinces*. Pictured Rocks is located within the Great Lakes section of the Central Lowlands physiographic province (Thornbury, 1965). Topographically, the park consists of three parts: the Pictured Rocks cliffs in the west, the Grand Sable Banks/Dunes in the east, and a 12-mile stretch of broad, sandy beach fringing the Beaver Lake basin in between (Figs. 1.2a, b). The Pictured Rocks cliffs vary from 50 to 200 feet above Lake Superior. Relief is moderate by midwestern standards, averaging 50 to 100 feet, but it reaches nearly 400 feet along the crest of Grand Sable Dunes, where some of the park's highest terrain is located only about one-half mile from the park's lowest point, Lake Superior. The national lakeshore's highest point is just east of Legion Lake near the junction of County Road H-58 and Little Beaver Lake Rd., where elevations exceed 1,060 feet. The park is bordered on the south by the hilly uplands of the Munising moraine, an accumulation of glacial debris formed along the retreating ice margin about 11,500 years ago, where elevations exceed 1,000 feet in a few places (Fig. 1.2b). In between the moraine and the coastline, the melting glaciers formed a stair-step series of meltwater stream terraces, becoming progressively lower toward the lake. In many areas, these features are truncated near the water's edge by abandoned shorelines assigned to the Nipissing and Algoma phases of the ancestral Great Lakes. A description of these features is included in the more detailed analysis of park terrain in chapters 3 and 4.

2

PRECAMBRIAN AND
PALEOZOIC BEDROCK

OVERVIEW AND PREVIOUS WORK

Systematic scientific study of the Lake Superior sandstones, which make up the
Pictured Rocks, begins with the work of Douglass Houghton (1841), Michigan's
first state geologist. Later, J. W. Foster and J. D. Whitney (1851), cruising the Lake
Superior shoreline in open boats, surveyed the region for the federal government
and described the sandstones in more detail (Whitney is best remembered today as
the namesake for California's Mount Whitney, the highest peak in the contiguous
United States). On the basis of mineral composition and fossil assemblages, Foster
and Whitney correlated Michigan's Lake Superior sandstones with the Cambrian
Potsdam sandstone of New York, suggesting a Cambrian age for the Michigan rocks.
A. C. Lane and A. E. Seaman (1907) later proposed the local name "Munising" for the
white sandstones in Pictured Rocks and "Jacobsville" for the underlying red strata.

Although a number of valuable studies were completed in succeeding years,
most of our knowledge regarding the Pictured Rocks sandstones comes from two
sources. The first is William Hamblin's dissertation, completed at the University of
Michigan and published as an official Michigan Department of Conservation booklet
in 1958. This research marks the first comprehensive analysis of sandstones within
the national lakeshore. Significant work on these rocks was long in coming because
the Munising sandstones are difficult to study. Many of the exposures are only
accessible by boat, and much of the rock is exposed in steep cliff faces high above lake
level. Over the course of several years, Hamblin picked and probed and hammered
and hacked his way up and down the coastline, often dangling precipitously from
ropes as he rappelled to areas of interest inaccessible from the beach. His efforts
were worth it, however, as the work remains the standard reference for the region,
even if some of his interpretations have been superseded by others. Charles Haddox
(1982) completed the second important study as part of his graduate studies at
the University of Wisconsin. Employing sophisticated advances in sedimentology

developed in the intervening 30 years, Haddox and his academic advisor, Robert Dott Jr. (Haddox and Dott, 1990), revised and reinterpreted many of Hamblin's conclusions, with some surprising results.

REGIONAL CORRELATION

Most researchers in the Pictured Rocks area continue to follow the chronology and nomenclature developed by Hamblin (1958) for the Munising and Au Train Formations, supplemented by new data and correlations with rocks from the northern Mississippi Valley (see Fig. 2.1; Thwaites, 1934; Oetking, 1951; Haddox and Dott, 1990; Miller and others, 2006). Others use different names for the uppermost units in the Pictured Rocks cliffs (Trempealeau Formation and Prairie du Chien Group) and assign the Munising Formation to the Middle (rather than Upper) Cambrian (for example, see Milstein, 1987). Until these discrepancies are sorted out, the situation will remain somewhat confusing to the unsuspecting reader. This book follows the terminology of Hamblin (1958), Haddox and Dott (1990), and Miller and others (2006), who have completed the most detailed studies of the Cambrian and Ordovician deposits in the region.

Hamblin (1958) divided the Munising Formation into three members: an unnamed basal conglomerate, the Chapel Rock sandstone, and the Miners Castle sandstone. Based on fossil assemblages, he assigned the Miners Castle Member to Late Cambrian time (see Fig. 1.7). The Chapel Rock sandstone lacks index

		Wisconsin	Northern Michigan	Cratonic sequences
ORDOVICIAN	MIDDLE	Platteville Fm.	Black River Fm.	Tippecanoe
	LOWER	St. Peter SS	Missing?	
		Prairie Du Chien	Au Train Fm.	
UPPER CAMBRIAN	TREMPEALEAUAN STAGE (Late Sunwaptan)	Jordan SS St. Lawrence Fm.	Missing?	Sauk III
	FRANCONIAN STAGE (Lower to middle Sunwaptan)	Tunnel City Group	Miners Castle Mb. / Munising Fm.	
	DRESBACHIAN STAGE (Late Marjuman and Steptoean)	Wonewoc Fm.	Chapel Rock Mb. / Basal Congl.	
		Eau Claire Fm. Mount Simon SS	Missing?	Sauk II
	PROTEROZOIC	Bayfield and Oronto Groups	Jacobsville SS	

FIGURE 2.1. Northern Michigan stratigraphy and correlations with Wisconsin and the Sauk/ Tippecanoe sequences according to Haddox and Dott (1990). Current stage names in parentheses.

18

fossils, making its exact age uncertain. Most workers (Hamblin, 1958; Driscoll, 1956; Ostrom and Slaughter, 1967) correlate it with the Wonewoc Formation of Wisconsin, indicating formation during the Late Cambrian (Fig. 2.1). The underlying Jacobsville sandstone also contains no fossils, making its age determination difficult. Evidence based on its magnetic properties (Roy and Robertson, 1978) indicates a late Precambrian age (late Keweenawan). Hamblin also recognized a sandstone of slightly different composition and structure capping the Munising strata, which he named the Au Train Formation. E. C. Guldzenopf (1967) and J. F. Miller and others (2006) firmly date the Au Train to the Early Ordovician period using fossil assemblages. Geologists are referred to figure 2.1 for details of correlations as recognized by Haddox and Dott (1990).

FIGURE 2.2. The Jacobsville Formation (lenticular facies) at Au Sable Point. Tape measure is extended 1 meter (3.3 feet).

JACOBSVILLE FORMATION

The Jacobsville Formation is the oldest rock exposed in Pictured Rocks National Lakeshore. It is named for the village of Jacobsville in Houghton County, Michigan, where it was quarried as Portage Redstone and was used in the construction of many important area buildings including the impressive Marquette County Courthouse in Marquette and Munising's City Hall. Its distinctive red color and widespread occurrence make it one of the most easily distinguishable rock units along Lake Superior's southern coast (Fig. 2.2). The formation consists of red and brown sandstones with thin seams of red clay

FIGURE 2.3. Leaching of hematite along fractures in the Jacobsville Formation east of Hurricane River Campground. Tape measure is extended 1 meter (3.3 feet).

shale. The red color is derived from an abundance of ferric iron oxide (hematite) eroded from Precambrian iron formations located in a highland source region to the south. In nearly every outcrop, however, the red color is famously mottled with white streaks, blotches, and circular spots caused by leaching along fractures, bedding planes, and cross-beds, or from chemical reduction of trace mineral impurities (Figure 2.3). Hamblin (1958) recognized four main facies (or subtypes) in the Jacobsville: a lenticular sandstone facies, a massive sandstone facies, a red siltstone facies, and a conglomeratic (gravel) facies. Haddox and Dott (1990) correlate the Jacobsville Formation with the Bayfield Group of sandstones exposed at Apostle Islands National Lakeshore in Wisconsin (Fig. 2.1).

Shoreline outcrops of the Jacobsville sandstone extend more than 150 miles along Lake Superior's southern shore from Bete de Grise Bay in Michigan's Keweenaw Peninsula to Munising in the east, broken only by the occasional sandy beach (see Fig. 1.10). Eastward from Munising to Beaver Lake, the Jacobsville is below, or sometimes only slightly above, the mean water level of Lake Superior. The most accessible outcrops in the park are at Hurricane River campground and Au Sable Point, described in the road log. Exposures also exist in steep cliffs along the western side of Grand Island just west of the park, and in the bluffs south of Grand Marais. Farther east, the formation is mostly covered by drift and can be followed in the subsurface using geophysical data and well logs all the way to Whitefish Point. A few small outcrops exist in the vicinity of Sault Ste. Marie (Hamblin, 1958). The Jacobsville is not present on the north shore of Lake Superior but likely forms much of the Lake Superior bottom, especially west of the Munising meridian. These rocks are a likely source for the red color in clays on the bottom of Lake Superior.

The Jacobsville lies unconformably on older Precambrian rocks of varying rock types, dominated by granite. The contact between the two is best viewed from the pier at Presque Isle Park in Marquette (50 miles west of Munising). This outcrop shows that the Jacobsville was deposited on an erosion surface of relatively high relief (at least four hundred feet), probably similar to that now observed in the Huron Mountains northwest of Marquette. Accordingly, the thickness of the Jacobsville varies markedly, pinching out completely where it laps onto Precambrian bedrock in the south, yet reaching over 1,500 feet at its type section in the eastern Keweenaw Peninsula. In the west, the formation is truncated by the Keweenaw Fault. Within the park, the top of the Jacobsville is truncated by the Basal Conglomerate Member of the Munising Formation. Ample evidence from Grand Island, just west of the park, indicates that the Jacobsville was tilted toward the northwest and eroded before deposition of the Munising Formation, indicating an unconformity or break in the geologic record of unknown duration.

Based on a detailed analysis of mineral composition, sedimentology, and geographical relationships, Hamblin (1958) concluded that the Jacobsville Formation was a mostly terrestrial (rather than marine) deposit formed by braided streams

FIGURE 2.4. Trough cross-bedding in the Jacobsville Formation east of Hurricane River Campground. Tape measure is oriented in the general direction of flow (northwest, or into the photo) and is extended 1 meter (3.3 feet).

flowing northward from an east–west trending highland source region to the south that he named the Northern Michigan Highland. As the highlands were uplifted, erosional debris was carried northward into a large basin located at approximately the present site of Lake Superior. As time passed, lakes formed within the basin, adding a predominantly lacustrine (or lake) component to deposition, which today is observable in the massive sandstone facies of the Jacobsville sandstone (mostly outside the park). Hamblin established the direction of stream flow by measuring the inclination of features called trough cross-beds at various locations. Trough cross-beds are composed of thin layers oriented at some angle to the main stratification in which the lower boundary is curved, in the manner of a stream channel cross-section, viewed at right angles to stream flow (Fig. 2.4). By plotting the direction of many trough cross-beds, the regional northward drainage pattern was established.

MUNISING FORMATION

The Upper Cambrian Munising Formation and its three distinct members comprise the majority of the Pictured Rocks within the National Lakeshore. These units persist throughout the outcrop, with only minor variations, and are distinguished based on mineral composition, grain size, sorting, and sedimentary structure (Hamblin, 1958). Some researchers include the basal conglomerate as part of the Chapel Rock Member, recognizing two rather than three members in the Munising Formation (Ostrom and Slaughter, 1967; Milstein, 1987). Hamblin (1958), however, ranked the conglomerate as a separate member, and Haddox and Dott (1990), recognizing the distinctiveness and persistence of the basal conglomerate, followed Hamblin, as does this book. The Basal Conglomerate Member (0 to 13 feet thick) is overlain by the Chapel Rock sandstone (40 to 60 feet thick), a resistant, well-sorted, medium-grained sandstone characterized by large-scale cross strata 3 to 6 feet thick. Next comes the Miners Castle sandstone, which forms the upper 140 feet of the Munising Formation and consists of poorly sorted, fine- to medium-grained sandstone with small-scale cross-bedding commonly less than a foot thick.

The Munising Formation dips gently to the southeast toward the center of the Michigan structural basin. East of Miners Castle, the formation also exhibits a localized western-component dip, so that the uppermost Miners Castle Member is exposed in western sections of the Pictured Rocks cliffs, with the Chapel Rock dominating the eastern end (Fig. 2.5).

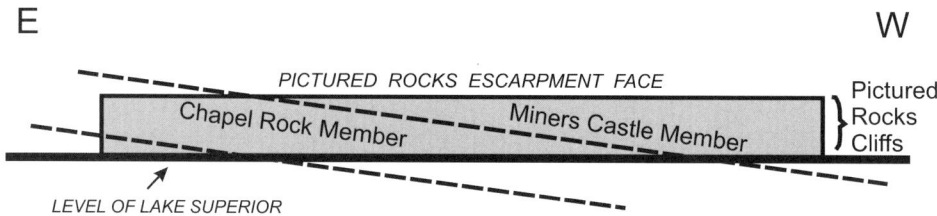

FIGURE 2.5. Highly simplified diagram showing the influence of the localized westward dip of the Munising Formation on outcrop patterns in the Pictured Rocks cliffs. View is southward. Vertical scale greatly exaggerated.

The Chapel Rock and the Miners Castle Members can be distinguished based on the size of the cross-bedding (large versus small), and by slope characteristics. The weaker Miners Castle sandstone often forms low- to medium-angle slopes, whereas the resistant Chapel Rock Member forms steep cliffs. The sharp contact between the two is often manifested as a distinct break in slope (Figs. 2.6, 2.7). Along much of the shoreline between Sand Point and Miners Castle, differential weathering and wave erosion along the contact have created a narrow bench or shelf on top of the Chapel Rock sandstone (Hamblin, 1958; Fig. 2.8).

The Pre-Munising Surface

The Jacobsville-Munising contact is clearly exposed in the Pictured Rocks escarpments (Fig. 2.9), but is best viewed along the cliffs of Grand Island, where it can be traced for several miles on both the east and west sides of the island. In places a distinct northerly dip of Jacobsville strata (4 to 5 degrees) can be observed in the cliffs, truncated by the sharp, nearly horizontal contact with the overlying Basal Conglomerate Member of the Munising Formation, which dips southeastward. This slight angular discordance (a type of unconformity) led Hamblin (1958) to conclude that the Jacobsville Formation was tilted and then eroded for an indeterminate amount of time, before deposition of the Munising sandstone was initiated. Haddox and Dott (1990), based on evidence observed along a contact at Grand Portal Point, concluded that the Jacobsville sandstone was already lithified (turned to stone) before the Munising sandstone was deposited.

Basal Conglomerate Member

Hamblin (1958) describes the Basal Conglomerate Member as an orthoquartzite (meaning "clean" or of high mineralogical purity) conglomerate containing well-rounded clasts (rock pieces) of quartz, quartzite, and chert, which make up nearly 90 percent of the conglomerate. Less than 1 percent of the constituent pebbles are of the Jacobsville Formation, a situation Hamblin ascribes to the mechanical instability of Jacobsville pebbles. Where coarse gravels predominate, little internal stratification is evident, but where significant amounts of sand are mixed in, distinctive high angle and planar cross-bedding indicate northward transport. No fossils have ever been found in this member, making its exact age difficult to determine. However, the

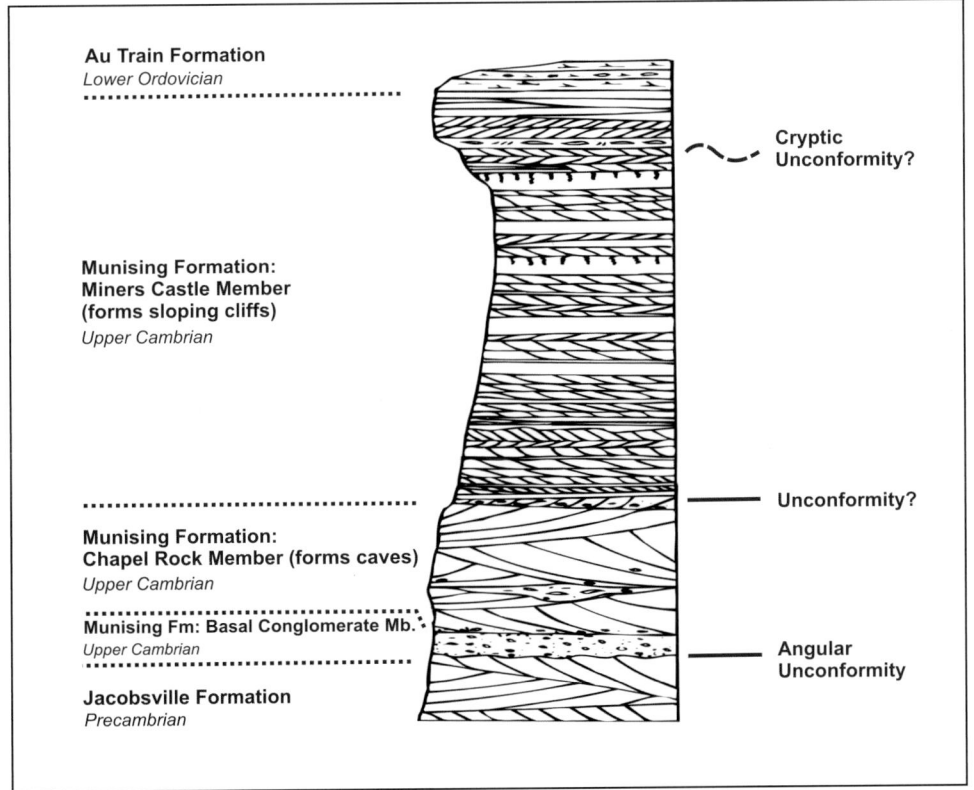

FIGURE 2.6. Bedrock profile of the Pictured Rocks, from Rosé (1997). (Drawing courtesy of National Park Service)

FIGURE 2.7. Bridal Veil Falls, a good place to view the slope characteristics of the Au Train Formation (top), the Miners Castle (middle), and the Chapel Rock Members of the Munising Formation (bottom) in the Pictured Rocks cliffs. Arrows show the approximate locations of the contacts. The entire section of the Miners Castle Member is exposed here and in nearby cliffs.

contact with the overlying Chapel Rock Member, firmly dated to the Late Cambrian, is gradational in some places, suggesting a similar, if slightly older, age for the basal conglomerate. The member is best observed by boat at Grand Portal Point (Fig. 2.10) and on the eastern side of Grand Island, as well as on land near the bottom of Sable Falls.

Based on composition, grain size, sorting, internal stratification, and geographical extent, Hamblin (1958, p. 76) interpreted the Basal Conglomerate Member as a gravel "blanket" deposited by transgressing seas over a Jacobsville surface of low relief. Haddox and Dott (1990), based on a detailed reexamination of the Munising Formation, and taking advantage of conceptual advances and techniques in sedimentology unavailable to Hamblin, reinterpreted the basal conglomerate as a low-sinuosity braided stream deposit, with a source area from the south or southeast. They state (1990, p. 702):

FIGURE 2.8. Bench developed at the contact (arrow) between the Chapel Rock (bottom) and the Miners Castle Members (top) of the Munising Formation at the east ledge of Miners Beach.

FIGURE 2.9. Abrupt contact (arrow) between the Jacobsville sandstone (bottom) and the Basal Conglomerate Member (middle gray unit) of the Munising Formation near Grand Portal Point. Note subtle cross-bedding in the conglomerate unit.

> A marine swashzone origin [such as that proposed by Hamblin] should have produced thin, continuous, well-sorted conglomerate beds and lenses . . . rather than the relatively thick, poorly sorted, massive, and cross bedded types of the basal Munising. . . . There seems little doubt that the initial deposition upon the pre-Munising unconformity was by fluvial processes.

This means that the basal conglomerate was likely deposited in a manner similar to the Jacobsville, with braided streams flowing northward from a highland to the south, although the center of the highland region may have shifted location somewhat during deposition.

FIGURE 2.10. The Basal Conglomerate Member of the Munising Formation near Grand Portal Point, with overlying high-angle cross-sets of the Chapel Rock Member.

Chapel Rock Member

This member is named for Chapel Rock, a conspicuous landmark located along the eastern end of the Pictured Rocks, where the unit is particularly well exposed (Fig. 2.11). The Chapel Rock Member is a well-sorted, medium-grained, unfossiliferous, large-scale cross-bedded and horizontally stratified sandstone with minor but widespread conglomerate, and interbedded black mudstone in its upper part. Sandstone within the member is composed almost exclusively of quartz, chert (microcrystalline quartz), and quartzite grains, with minor amounts of feldspar (Hamblin, 1958). Excellent exposures of this unit are found along the entire extent of the Pictured Rocks and at Sable Falls. Because of the Munising Formation's localized southwest dip (Fig. 2.5), the Chapel Rock Member is especially well exposed along the Pictured Rocks east of Mosquito River, where it forms virtually the entire section. Farther west, only the uppermost 10 to 15 feet is exposed above lake level between Miners Castle and Munising (Figs 2.7, 2.8). Here, the cliffs are formed mostly upon the Miners Castle Member.

FIGURE 2.11. Chapel Rock, the type locality for the Chapel Rock Member of the Munising Formation (Hamblin, 1958).

The Chapel Rock Member is approximately 40 to 60 feet thick within the lakeshore and may thin laterally (Hamblin, 1958). Its areal extent to the east, south, and west is uncertain since few exposures exist outside of the park. Eastward, it is exposed in the lower falls of the Tahquamenon River and at Encampment d'Ours Island in the St. Marys River. Westward, it is found on the flanks of Limestone Mountain in Houghton County, but its southern extent is still a matter of conjecture.

Haddox and Dott (1990) recognized four principal facies (sediment types representing different depositional environments) within the Chapel Rock Member: (1) cross-bedded sandstone, (2) horizontally stratified sandstone, (3) channelized conglomerate and related breccias (breccias are similar to conglomerate but the clasts have angular rather than rounded corners), and (4) laminated (or layered) mudstones. Large-scale trough cross-bedding is the most conspicuous sedimentary structure in the Chapel Rock Member (Fig. 2.12 a, b). Trough widths range from 3 to 600 feet, with an average of approximately 30 feet. The troughs are ubiquitous and one of the most useful features in distinguishing the Chapel Rock from other units. Very shallowly dipping and horizontally stratified sandstones are also common (Fig. 2.13), along with mud cracks, ripple marks, and fossil tracks. These different facies each represent a different environment of deposition and attest to the wide variability of depositional processes at the time of formation.

FIGURE 2.12 a, b. Large, high-angle cross-bedding and reactivation surfaces in the Chapel Rock Member. Top photo taken along the cliffs west of Miners Castle. Bottom photo is from the east ledge of Miners Beach. Tape measure (arrow) in bottom photo is extended 1 meter (3.3 feet).

FIGURE 2.13. Low angle cross-strata in the Chapel Rock Member at Chapel Rock.

Mud cracks tend to be associated with the mudstones and are best exposed just below the contact with the overlying Miners Castle Member (Fig. 2.14). Hamblin (1958) reported polygonal mud crack patterns ranging from 4 to 18 inches across (Figs. 2.15, 2.16). The depth of the cracks is limited by the thinness of the beds and rarely exceeds 6 inches. Polygonal cracks filled with sand are common (Fig. 2.16). Accessible exposures are located at the east end of Miners Beach and at Mosquito Beach.

FIGURE 2.14. Mudstones exposed at the top of the Chapel Rock Member, west end Miners Beach. Tape measure is extended 1 meter (3.3 feet).

Ripple marks are spectacularly displayed within horizontally stratified sandstones at Mosquito Beach (Fig. 2.17), although they can be found throughout the park. Geologists value well-preserved ripple marks because they can help indicate the local environment at the time of deposition. Ripple marks that are symmetrical in profile indicate oscillating current, as might be expected from waves along a shoreline or in a tidal flat, and can indicate the trend of the ancient shoreline. Asymmetrical ripples form in strong currents and indicate the direction of sediment transport. Both types of ripple marks can be identified within Pictured Rocks National Lakeshore (Haddox and Dott, 1990). Some ripple marks exhibit superimposed mud cracks (see Fig. A.10), indicating drying and shrinking of overlying clay layers during intermittent exposure to the atmosphere (Haddox and Dott, 1990).

Sand concretions (spherical accumulations of calcite within the sandstone, resembling rounded pebbles) are common in the Chapel Rock Member where extensive fracturing has increased permeability and allowed the secondary diffusion and accumulation of calcite. Concretions average 1 to 2 inches in diameter, with a few exceeding 10 inches in their longest dimension.

Origin of the Chapel Rock Member

Because the Chapel Rock Member exhibits an unusual assemblage of distinct bedforms representing a wide range of depositional environments, its interpretation has been especially challenging. Hamblin (1958) interpreted the basal conglomerate and overlying Chapel Rock Member as two components of a classic transgressive-regressive marine sequence. In other words, he saw the Chapel Rock unit essentially as

FIGURE 2.15. Mud cracks exposed at the west end of Mosquito Beach in the Chapel Rock Member. Tape measure is extended 10 centimeters (3.9 inches).

FIGURE 2.16. Sand-filled mud cracks formed in shale deposited with low-angle strata in the Chapel Rock Member, west end of Mosquito Beach. Tape measure is extended 10 centimeters (3.9 inches).

FIGURE 2.17 a, b. Asymmetrical ripple marks at the east end of Mosquito Beach. Note parallel tool marks (produced as objects such as shells or pebbles skip or drag across a sedimentary surface) or possible trace fossils in right photograph. Tape measure extended 1 meter (3.3 feet) in left photo, 10 centimeters (3.9 inches) in photo at right. Paleocurrent direction is right to left in a, front to back in b.

a marine deposit representing a rising sea level that peaked and began to fall with deposition of the Chapel Rock Member. He interpreted the interbedded shale lenses within the top of the member as evidence of regression:

> It is significant to note that the zone containing the mud cracks is the uppermost unit of the Chapel Rock member and in many places it is in direct contact with basal units of the Miner's Castle member. This indicates that the upper Chapel Rock member was deposited in a shallow-water environment which was repeatedly exposed to subaerial [exposed to the open air] conditions. This might be interpreted as a regressive phase of the Chapel Rock sea. (Hamblin, 1958, p. 86–87)

Haddox and Dott (1990), in their reexamination of the Munising sandstone, presented a much more complicated interpretation (based on conceptual advances unavailable to Hamblin), recognizing both terrestrial and marine influences. They state (1990, p. 697):

The . . . Chapel Rock Member has unfossiliferous, large-scale cross bedded sandstones with minor but widespread conglomerate interpreted as braided fan-delta distributary deposits and associated horizontally-stratified sandstone with wave ripples, polygonally-cracked shales, and trace fossils, interpreted as interdistributary deposits. Both of these latter facies are punctuated by unsorted, unstratified, channelized conglomerate and intraclast breccia, which are interpreted as river-flood avulsion deposits. Although fluvially-dominated, the depositional system also had wave and probable tidal influences as well as eolian reworking of fluvial sands. Thus a shoreline setting adjacent to highlands centered over the northern Michigan-Wisconsin border is indicated.

To those unfamiliar with geologic terminology, only the last sentence will make much sense. The essence of the argument, however, is that marine influences may have been less important in forming the Chapel Rock Member than originally assumed, with a greater emphasis on fluvio-deltaic (stream deltas) and eolian (wind generated) deposition. This interpretation builds upon the findings of Hamblin and provides a much more detailed understanding of the geologic history, which will be presented shortly.

Miners Castle Member

The Miners Castle Member forms the upper 140 feet of the Munising Formation (Figs 2.6. 2.7), and consists of light gray to white, poorly sorted, somewhat friable silty-shaley quartz sandstone, containing small cross-beds averaging 4 to 6 inches thick. Thin lenses of blue shale separate cross strata in the lower part of the unit, giving way to medium- and well-sorted upper units of mostly sandstone. Quartz grains make up more than 95 percent of the unit, with minor amounts of feldspar. Garnet is an important heavy mineral constituent and is used to differentiate the Miners Castle from the Jacobsville Formation and Chapel Rock Member, whose heavy-mineral fractions are dominated by zircon. The Miners Castle Member is very poorly sorted, with grain sizes varying from gravels to silt and clay. Mud cracks, ripple marks, and concretions are also present (Hamblin, 1958). The lower 90 feet of the Miners Castle Member is dominated by small-scale cross-stratification with alternating thin lenses of bluish-green shale. The upper part of the unit exhibits extensive horizontal bedding 2 to 8 inches thick that can be traced laterally for many miles along the lakeshore cliffs. It is more resistant than the lower part of the member and readily observed at Bridal Veil Falls (Fig. 2.7) where it forms a white vertical cliff near the top of the section.

The Miners Castle Member is best observed at its type section at Miners Castle (Fig. 2.18) and along the Lake Superior shoreline from Sand Point to Miners Castle, where the entire section is exposed (Fig. 2.7). East of Miners Castle, the Chapel Rock Member predominates due to the western dip of the strata mentioned previously (Fig. 2.5). It is best differentiated from the underlying Chapel Rock Member by the predominance of small-scale (versus large-scale) cross-bedding, tendency to form slopes rather than cliffs, and dominance of garnet over zircon in the heavy-mineral fraction. Readers seeking a more detailed description are referred to Hamblin (1958), Haddox (1982), and Haddox and Dott (1990).

Hamblin's (1958) analysis of the plunge directions of trough cross-bedded units within the Miners Castle Member showed a remarkably consistent regional slope toward the west–southwest, indicating that its source area was to the northeast in Canada. This situation contrasted sharply with plunge directions from the Jacobsville and Chapel Rock units, which indicated northward drainage from a source area in the Northern Michigan Highlands to the south. According to Hamblin (1958, p. 109):

FIGURE 2.18. Miners Castle, the type section for the Miners Castle Member (Hamblin, 1958) from the lake. Contact between Chapel Rock (bottom) and Miners Castle Members (top) is clearly observed about 1 meter above the water level.

This marked change in the direction of regional slope during the deposition of the Chapel Rock and Miner's Castle members further indicates that the two members are separated by an unconformity. The Wisconsin arch and the Northern Michigan Highland, which were prominent source areas during Jacobsville and Chapel Rock time, were eroded down and almost completely covered with the Chapel Rock sediments. It is probable that regional tilting caused a regression of the Chapel Rock sea so that by the beginning of Miner's Castle time the regional slope was to the southwest. The transgression of the Miner's Castle sea was from the southwest, across the eroded Wisconsin arch and Northern Michigan Highland, but the major source area lay farther to the northeast in Canada. This change in principal source area clearly explains the change in heavy mineral suite from a high zircon–low garnet in the Jacobsville formation and Chapel Rock member to a high garnet–low zircon in the Miner's Castle member.

Simply put, these relationships mean that a significant amount of erosion and regional tilting, perhaps lasting several million years, occurred between the deposition of the Chapel Rock and Miners Castle Members. Haddox and Dott (1990, p. 709–710), however, questioned this interpretation and attributed changes in the source area to changes in the way sediments were laid down in a deepening ocean. Their interpretation does not necessarily require a significant period of erosion or a complicated sequence of shoreline advances or retreats. The significance of these proposals will be reviewed in the summary of the geologic history that follows.

AU TRAIN FORMATION

The Au Train Formation (Grabau, 1906; Hamblin, 1958) of Early Ordovician age lies unconformably above the Munising Formation and is the youngest Paleozoic rock formation exposed in the park (see Fig. 1.7). It consists primarily of carbonate-rich sandstone, sandy dolomite, and beds of pure dolomite (dolomite is a carbonate rock similar to limestone in many respects, but with higher amounts of magnesium). Hamblin designated Au Train Falls, 12 miles southwest of Munising, as the representative section (Fig. 2.19) and divided the formation into a lower (100-feet-thick) and upper (230-feet-thick) member based on the presence of the mineral glauconite in the lower member. The upper half of the formation is known only from drill holes located south of the park. The lower part of the formation represents a near shore shallow marine environment associated with a widely recognized Early Ordovician marine transgression (Sauk III, see below) that covered large parts of interior North America (Fig. 2.1) (Miller and others, 2006).

FIGURE 2.19. The lower member of the Au Train Formation near its representative section at Au Train Falls (Hamblin, 1958), 12 miles southwest of Munising.

The Au Train Formation forms the cap rock for the Pictured Rocks cliffs and is easily distinguishable as a light-gray to bluish-gray to brown rock unit about 10 to 30 feet thick along the top of the cliffs between Miners Castle and Sand Point (Figs. 2.7, 2.20). Here, the more resistant dolomitic sandstones at the base of the unit form steep cliffs and often stand in stark contrast to the more gradual slopes and alcoves of the underlying and weaker Miners Castle Member of the Munising Formation. As figure 2.6 shows, however, the actual lithologic contact, where quartz-rich sands of the Munising Formation give way to predominantly dolomitic sands above, is located

FIGURE 2.20. The Au Train Formation forms the cap rock for the Pictured Rocks cliffs and is a distinctive light gray rock unit (arrow) about 10- to 30-feet thick along the top of the cliffs just east of Miners Beach.

higher up within the cliff (Rosé, 1997). The resistant character of the Au Train is responsible for maintaining many of the waterfalls in Alger and Marquette Counties, including Munising, Laughing Whitefish, and Au Train Falls.

FOSSILS IN PICTURED ROCKS NATIONAL LAKESHORE

Fossils are relatively rare in rocks of the National Lakeshore. No recognizable fossils exist in the Jacobsville Formation or in the Basal Conglomerate Member of the Munising Formation. Trace fossils (meaning a track, trail, or burrow left behind by an organism) occur within mudstone and sandstone facies of the Chapel Rock Member within the park, but are exceedingly difficult to identify except by experts. Nearly all involve tracks left behind by crawling or furrowing organisms, mostly trilobites (Cruziana and Rusophycus; Haddox and Dott, 1990; Fig. 2.21). Trilobites are an extinct group of marine organisms in the phylum Arthropoda that are related to modern spiders, insects, and crustaceans. Body fossils (meaning casts or remains of the organism itself rather than its tracks or imprints) are found within the Miners

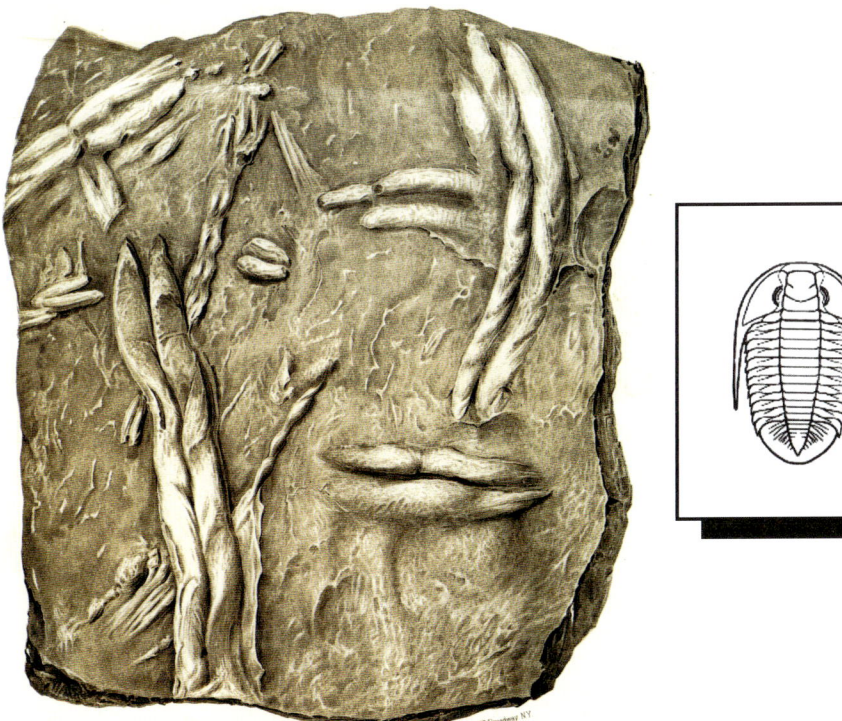

FIGURE 2.21. Trilobite furrowing (*Cruziana*—the longer paired depressions) and resting tracks (*Rusophycus*) in the Chapel Rock Member. The tracks were made in muds and then preserved as casts on the bottom of sandstones deposited on top of the mud (from Foster and Whitney, 1851). Inset shows a trilobite of the genera *Saukia*, a close relative of those found in the Munising Formation.

Castle Member and include the trilobites Prosaukia, Briscoia, Idioniesus, and Idahoia (Fig. 2.22), the brachiopod Lingulepis (brachiopods are bivalved shelled organisms that externally resemble modern clams), and the cephalopod Michelinoceras (cephalopods are marine mollusks whose modern forms include octopuses and squids). Other trace fossils reported by Haddox (1982) include Planolites and Skolithos (tubelike lineations in sandstones thought to represent the burrows of marine wormlike organisms). He also reports that many massive sandstone beds near the top of the Miners Castle Member are completely bioturbated (meaning that they have been churned and thoroughly mixed by organisms; such characteristics are difficult to recognize in outcrops except by experts).

FIGURE 2.22. Diorama showing marine life during the Early Paleozoic Era, about 400 million years ago. Ocean life exploded during the 54 million years of the Cambrian, creating an enormous variety of genera and species. Note trilobites in center foreground. (Courtesy of National Park Service)

The Au Train Formation contains Early Ordovician age trilobites, brachiopods, cephalopods, gastropods (a type of mollusk represented by modern snails), hyoliths (a type of mollusk), and conodonts (Fig. 2.23). Conodonts are preserved as small, often microscopic, sawtoothlike fossils that are extremely useful in correlating geologic units. Parts of two trilobite fossils, along with abundant conodont assemblages, were collected from the Au Train Formation near Sand Point by geologist Bob Rosé. Analysis of these specimens by Dr. Rosé and his colleagues (Miller and others, 2006), along with a review of previous work, indicated that an important assemblage of fossils associated with what geologists call the

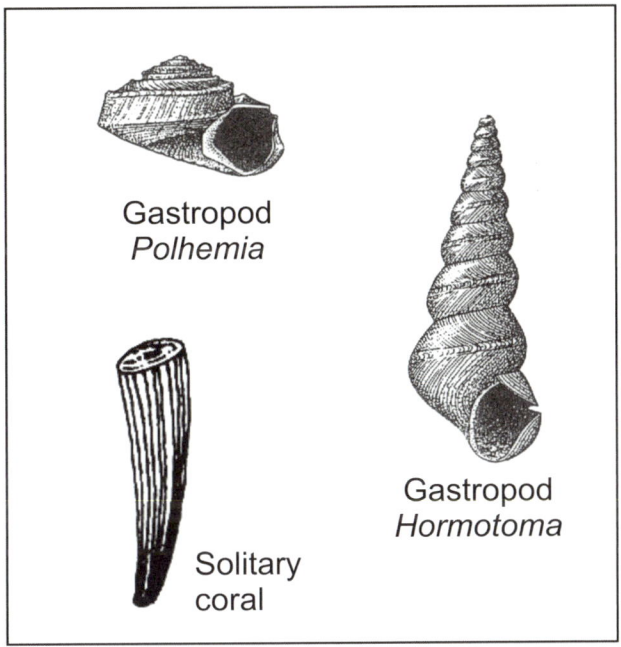

FIGURE 2.23. **The marine Au Train Formation contains fossilized corals and gastropods (shown here), along with brachiopods and trilobites.**

Trempealeauan Stage (Figure 2.1) was missing from Pictured Rocks strata, indicating that an unconformity exists at the base of the Au Train Formation. Curiously, they also found that the upper part of the Miners Castle Member likely dates to the Early Ordovician rather than the Late Cambrian, meaning that the unconformity exists somewhere in the upper part of the Miners Castle Member rather than at the contact with the overlying Au Train Formation. This situation illustrates the discrepancies and confusion that can occur when divisions in the geologic record are made on the basis of changes in rock type versus changes in fossil assemblages. It also means that the Munising Formation, long assigned to the Cambrian period, is actually both Cambrian and Ordovician in age.

GEOLOGIC HISTORY: SEQUENCE OF EVENTS

What were the events that led to the formation of the Pictured Rocks sandstones? Fifty years ago the answers to such questions were elusive and fraught with uncertainties. Since then, plate tectonics has revolutionized our view of the Earth and has allowed for a much more satisfying and comprehensive understanding. In this chapter we'll confine our discussion to events related to formation of the Precambrian and Paleozoic bedrock. Chapters 3, 4, and 5 will address Pleistocene and Holocene events, including the formation of Lake Superior.

The geologic history of Pictured Rocks National Lakeshore begins with the deposition of the Jacobsville sandstone as early as a billion years ago during the later part of the Proterozoic Eon (see Fig. 1.6) (Kalliokoski, 1982; Catacosinos and Daniels, 1991). By this time the Lake Superior region had undergone a long and complicated geologic history lasting more than 3 billion years, culminating in the formation of the continental landmass Laurentia, the forerunner of modern North America. By about 1.1 billion years ago, Laurentia began to split apart along a three-pronged rift or crack in the Earth's crust (like today's Rift Valley of east Africa), centered on the Lake Superior region (Fig. 2.24) (NICE Working Group, 2007; Holm and others, 2007; Schulz and Cannon, 2007). Rifts occur where the Earth's crust is stretched and broken, resulting in a linear depression. The Lake Superior rift axis not only provided a lowland trap for sandy sediments washed in from the adjacent highlands but also served as the source for enormous outpourings of basaltic lavas. The resulting intermingled lavas, sandstones, and conglomerates make up what geologists call the

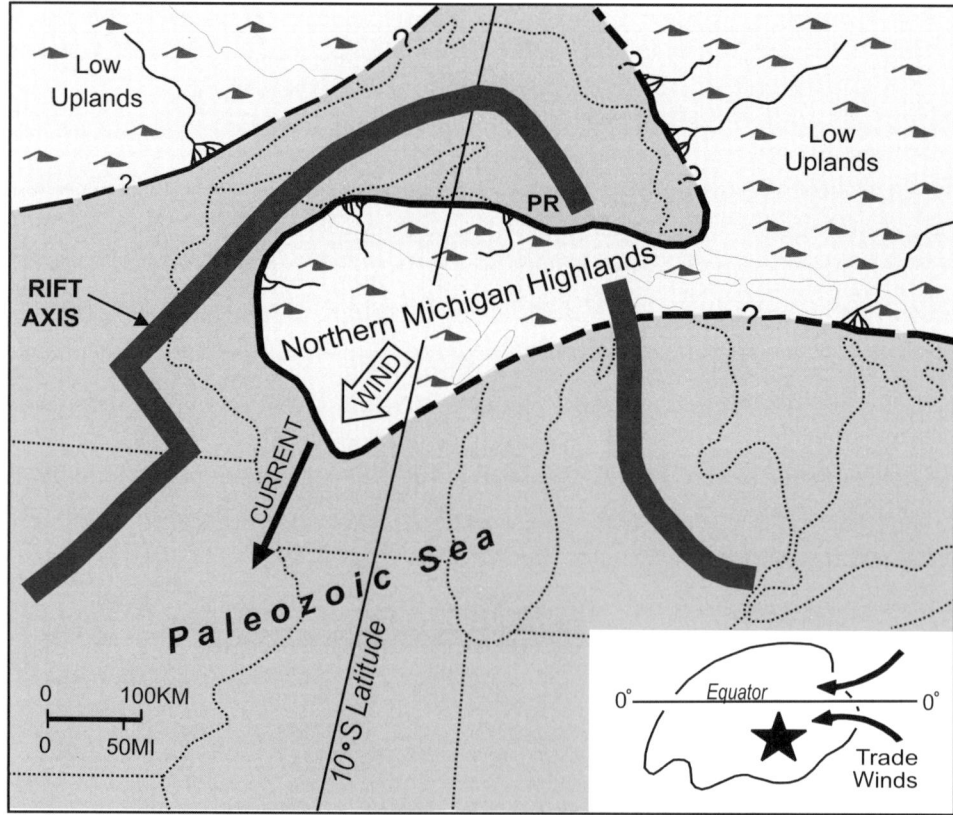

FIGURE 2.24. Early Late Cambrian paleogeography of the Great Lakes region during deposition of the Chapel Rock Member, as interpreted by Haddox and Dott (1990). Rift axis shown as thick black line. Inset shows the Cambrian paleolatitude of North America with the Pictured Rocks area indicated by a star.

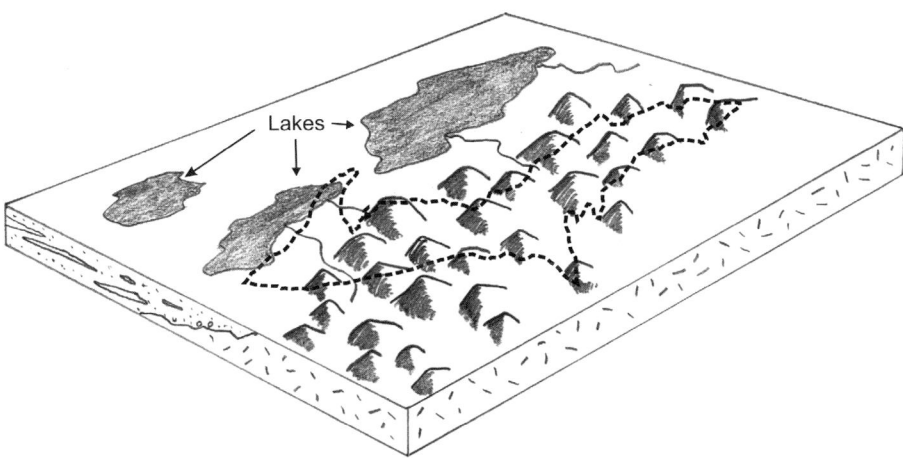

FIGURE 2.25. Paleogeography of Northern Michigan during deposition of the Jacobsville Formation. The Northern Michigan Highlands extended eastward from northern Wisconsin through the central part of Michigan's Northern Peninsula and into Canada (after Hamblin, 1958).

Keweenawan series. These rocks are the source for the enormously valuable copper deposits of Michigan's Keweenaw Peninsula. The downward drop of crustal blocks along the rift axis was likely mirrored by an upward warping along its edges, forming a distinct highland centered on what today is Michigan's Upper Peninsula. This Northern Michigan Highland persisted in different forms and for different reasons for millions of years, serving as the source area for sediments that would eventually form the Jacobsville and Munising Formations. That the paleogeography was perhaps more complicated than this is indicated by the fact that the eastern extension of the rift axis passes nearly directly beneath Pictured Rocks National Lakeshore (Fig. 2.24).

The titanic forces working to tear Laurentia apart eventually subsided and rifting ceased. The lowland created by this rifting, however, persisted in different forms for millions of years afterward, well into Jacobsville time. Thus, at the time of Jacobsville deposition, a moderate relief Northern Michigan Highland, similar in some respects to the modern Appalachians, fronted an extensive lowland centered on what is now Lake Superior (Fig. 2.25). Braided streams washed debris eroded from the highlands downslope (northward in modern direction) into the basin. Distinct lacustrine facies within the Jacobsville indicate that small lakes were also present in the basin, although stream deposition was by far the dominant process (Hamblin, 1958). As deposition continued, the Northern Michigan Highlands were eventually buried, or nearly so, in their own debris.

This landscape resembled nothing observed on Earth today. Land plants and animals had yet to evolve and only barren rock mantled with loose debris would have been exposed at the surface. Indeed, except for the streams, the scene would have resembled a Martian landscape. Iron-bearing minerals in the Jacobsville deposits reacted with the humid environment and were altered to the distinctive red color observed in the Jacobsville Formation today. Through natural processes these deposits were eventually lithified, or turned to stone, probably well before the Proterozoic Eon ended.

A significant unconformity or break in the record separates the Jacobsville Formation from the overlying Munising Formation. During an extensive period of weathering and erosion lasting millions of years, Jacobsville rocks were tilted slightly to the northwest and then eroded to a low relief surface. Tilting may have been associated with renewed uplift of the Northern Michigan Highlands, which initiated the next round of deposition associated with the Munising Formation during the early part of the Late Cambrian.

By this time, Laurentia was astride the equator and the Pictured Rocks area was located about 10 degrees south of the equator (Fig. 2.24; Haddox and Dott, 1990). Because much of Laurentia was close to sea level and of low relief, it was especially susceptible to flooding of its interior by surrounding seas. A useful analogy is to imagine Laurentia as a waterlogged raft in a Paleozoic sea, bobbing and sinking with the waves, intermittently dry on one side as the other sinks below the surface. The rock record of North America clearly reveals the pattern and timing of these marine transgressions (flooding of the land) and regressions (draining of the ocean from the land).

Today we attribute these changes in sea level (at least in part) to the relative intensity of plate tectonics processes. Increased uplift along rift axes caused by rising plumes of magma displaced great volumes of seawater, causing global sea levels to rise and flooding low-relief continental margins and interiors. Decreased activity led to less uplift and sea levels fell. Geologists refer to these cycles of marine transgression and regression as cratonic sequences (Fig. 2.1) and recognize four such sequences in the Paleozoic (from oldest to youngest): Sauk, Tippecanoe, Kaskaskia, and Absoraka (Sloss, 1963; Levin, 1999). Paleozoic rocks in Pictured Rocks National Lakeshore are associated with the earliest of these transitions, the Sauk, which peaked in the Late Cambrian. By this time, these seas had inundated large parts of interior North America and were encroaching upon the Northern Michigan Highlands from the northwest.

In contrast to the sterile Precambrian landscape, life in the oceans exploded during the 54 million years of the Cambrian period, creating an enormous variety of genera and species (Fig. 2.22). Although many life forms eventually became extinct, the basic template of life laid down in the Cambrian remains the basis for much of life on Earth.

As these seas slowly advanced, swift-flowing braided streams were depositing extensive aprons of sand and gravel along the northern ramparts of the Northern Michigan Highland (Fig. 2.26). These deposits today form the basal conglomerate of the Munising Formation. With the continued advance of the sea, the depositional environment shifted to delta, beach, dune, and tidal environments. Because stream gravels are interbedded with beds containing marine fossils and wave ripples, Haddox and Dott (1990, p. 713) favored a "coastal setting with mixed fluvial [stream] and marine influences" as one might encounter in a steep-gradient marine deltaic system.

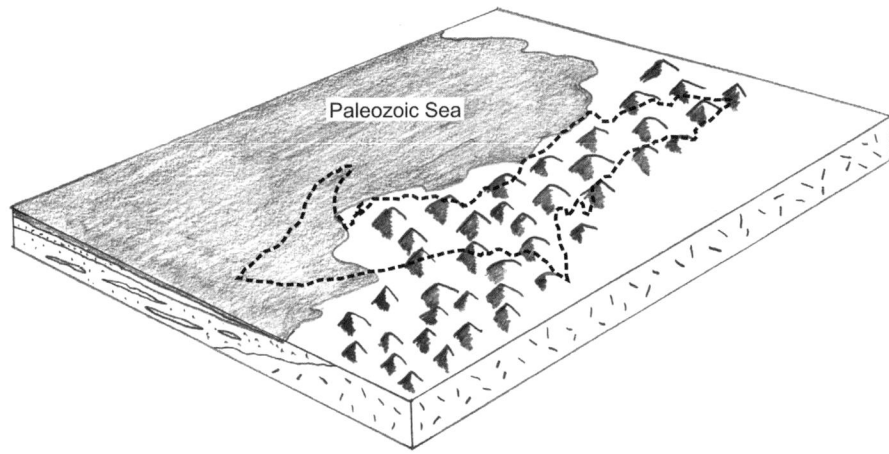

FIGURE 2.26. Paleogeography of Northern Michigan during deposition of the Chapel Rock Member of the Munising Formation. The Northern Michigan Highland persisted despite having been covered with Jacobsville sediments (after Hamblin, 1958). Note similarity to Haddox and Dott (1990) in Fig. 2.24.

Continued advance of the Paleozoic seas eventually submerged most of the Northern Michigan Highlands, shifting the source area for sediments to the northeast in Canada (Fig. 2.27), at least according to Hamblin (1958), and initiating deposition of the Miners Castle Member. According to Haddox and Dott (1990), these rocks record continuing encroachment of the Paleozoic seas, with deposition shifting from a near-shore marine environment containing silt and mud to a shallow marine shelf setting with strong currents.

After deposition of the Miners Castle Member, the region may have been uplifted and eroded, based on the presence of an unconformity between the Late Cambrian Miners Castle Member and the overlying Early Ordovician Au Train Formation (Dorr and Eschman, 1970). The Paleozoic seas had returned by the Early Ordovician, however, as evidenced by mixed carbonate rocks and sandstones in the lower part of the Au Train Formation indicating a near-shore shallow marine environment.

Thus, the rocks of the region represent a complicated sequence of events, initially dominated by stream processes (Jacobsville sandstone), followed by mixed stream, deltaic, eolian, and marine conditions (Chapel Rock sandstone), culminating in a predominantly marine environment by transgressing Paleozoic seas (Miners Castle Member and Au Train Formation).

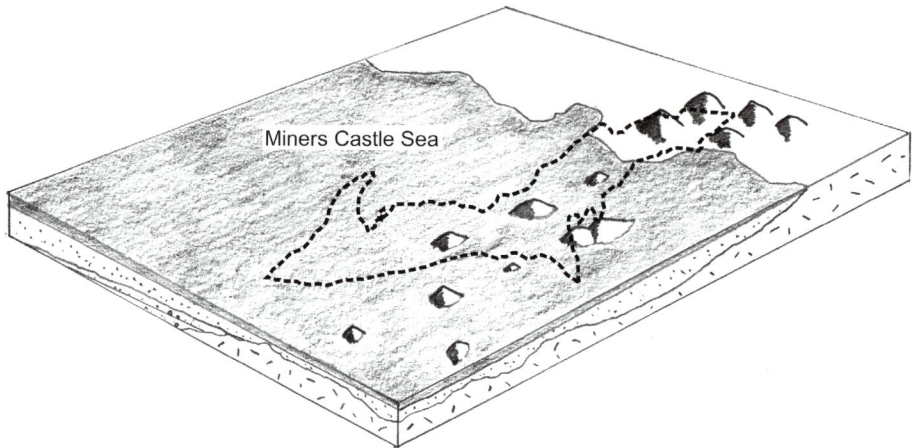

FIGURE 2.27. Paleogeography of Northern Michigan during deposition of the Miners Castle Member of the Munising Formation (after Hamblin, 1958). In this interpretation, the Northern Michigan Highlands have been eroded and buried in their own debris and then flooded by a Paleozoic sea advancing from the southwest.

THE LOST INTERVAL

Bedrock strata lying above the Au Train Formation are missing in Pictured Rocks National Lakeshore, and the geologic record is completely blank from the Late Ordovician to the very latest part of the Pleistocene, a time span encompassing more than 400 million years. Events from this "lost interval," at least the first part of it, must therefore be deduced from rocks in southern Michigan. Here, marine deposition associated with Paleozoic seas continued nearly uninterrupted throughout much of the remainder of the Paleozoic era, forming the enormous volume of sediments associated with the Michigan structural basin. These deposits likely extended northward into the Upper Peninsula and covered the Au Train but were later removed by erosion.

By the end of the Paleozoic, the downwarping that had been so long associated with the Michigan basin ceased, the seas drained away, and the region began an uninterrupted 280 million years of gentle uplift and erosion lasting well into the late Cenozoic. By the end of the Pliocene, differential weathering had probably produced a classic ridge and valley landscape of outfacing cuestas developed upon the upturned edges of Michigan basin strata. As discussed in chapter 1, the northward-facing Cambro-Ordovician escarpment associated with the Munising/Au Train

Formations represents the outermost of these cuestas (see Fig. 1.9), although its location during the Pliocene was probably farther north than its present position. At the same time, a major drainage network flowing toward the northeast had probably developed upon weaker strata lying between the escarpments (Spencer, 1891; Hough, 1958; Kincare and Larson, 2009). Thus, by the late Pliocene, all the major bedrock and structural elements reflected in the modern landscape were in place. Now, it was the glacier's turn.

3

ANCIENT LAKES AND
RELICT SHORELINES

The idea that Michigan and much of northeastern North America were buried beneath a vast continental ice sheet a mile thick is difficult for most people to imagine. Yet the evidence for continental glaciation, compiled and corroborated by hundreds of scientists over the past 150 years, is irrefutable. Their work documents not only the complicated interplay of glacier advances and ice marginal retreats but also the growth and extinction of a multitude of vast glacial lakes that formed along the fluctuating ice margin as the deep depressions of the Great Lakes were uncovered. Understanding the lake sequence is crucial to a thorough appreciation of the region's geologic history, yet it is exceedingly difficult to comprehend, even by the well initiated. Accordingly, this chapter will attempt to provide a cogent, although simplified, overview of present concepts regarding these ancient ancestors of the modern Great Lakes, giving special emphasis to the abandoned shoreline features observable in Pictured Rocks National Lakeshore. Chapters 4 and 5 then incorporate this information to provide a detailed review of Pleistocene and Holocene events in the region, completing the geologic history begun in chapter 2.

INTRODUCTION

Ever since the famous early promoter of the glacial theory, Louis Agassiz, first pondered the craggy outlines of Lake Superior more than a century ago, the abandoned shorelines of the Great Lakes region have intrigued geologists. Early investigators rightly attributed these features to periods of higher lake levels associated with a melting continental glacier during the Ice Age, but only recently has a coherent picture of the complicated series of glacial lakes and the strands they left behind begun to emerge.

Relict (or "raised") shorelines are common features along the Great Lakes. By "relict," we mean landforms such as beach ridges, wave-cut bluffs, shoreline caves, sand spits, and stacks that are now abandoned and higher than the modern shoreline. Some of these features are now located well inland and can be traced for considerable distances. To the casual visitor, however, their formation and significance are difficult to grasp because of the enormous size of the lakes involved and our unfamiliarity with the mechanisms that formed them. Many visitors also confuse the large-scale changes in lake levels from the Ice Age with the much smaller variations caused by changes in temperature, precipitation, and runoff that characterize the modern Great Lakes from year to year. The latter rarely exceed 3 feet, whereas the former reach into the hundreds of feet. Simplifying the science is not easy, but by understanding just a few major principles, most of the intricacies can be easily understood.

GETTING STARTED

We'll keep terminology to a minimum in this chapter, but a few terms are essential. For our purposes, basin refers to the elongated depressions in the Earth's crust that hold the present Great Lakes. For example, when we refer to the Superior basin, we mean "the hole in the ground" (to put it perhaps too simply) that contains modern Lake Superior. The modern lake, however, is only the latest in a number of water bodies that have occupied this particular lowland. Because these lakes varied in size, elevation, configuration, and in the location of their outlets, they've been given different names, such as Lake Minong and Lake Duluth, yet all formed in the Lake Superior basin. Lake phase refers to the particular time period during which a lake existed. The Nipissing phase of the Great Lakes, for example, occurred between about 4,000 and 6,000 years ago. Such time-dependent terms allow geologists to distinguish between the lake itself and the time during which that water body existed.

Geologists can determine the ages of glaciations and ancient shorelines using radiocarbon dating. This technique is based on the assumption that all living things contain a small proportion of the radioactive isotope 14C in their tissue. 14C is inherently unstable and will decay into the more stable by-product, 14N, at a known rate. This rate, called the half-life of 14C, is the time it takes for half of the 14C present to decay into 14N, about 5,568 years. When an organism is alive, the amount of 14C is continually replenished in tissue from the surrounding environment through respiration and intake of nutrients. However, upon death, the 14C begins to break down at a known rate, acting as a natural clock. If a lucky geologist later stumbles upon the preserved remains of the organism, the ratio of 14C to the total carbon in the sample can be measured and an age determined. In the Great Lakes region, wood is the material most often dated by geologists. Advancing glaciers often buried forests under thick mantles of glacial sediment (called till), and the wood can be sampled and dated to determine the time of burial. Along ancient shorelines, wood and other organic material often filled the low swales between beach ridges and were covered by sand dunes and preserved.

The production of 14C in the environment varies somewhat over time, however, and radiocarbon dates must be corrected in order to give true calendar dates. For example, the Marquette ice advance discussed in this chapter is dated through radiocarbon means to 10,025 ± 100 years B.P. (Before Present) (Lowell and others, 1999), a date that corresponds to 11,526 ± 219 calendar years (Fairbanks and others, 2005). This situation can make a detailed discussion of glacial events somewhat confusing to the general reader because Quaternary scientists often report and discuss their age estimates in the scientific literature as uncorrected dates, which are listed as "radiocarbon yrs. B.P." or "yrs. B.P." For ease of discussion, this book uses calendar years when discussing Pleistocene and Holocene events.

More recently, a new technique called Optically Stimulated Luminescence (OSL) (Aitken, 1998) allows geologists to actually determine when a geologic surface was last exposed to the sun's rays. The details are beyond the scope of this chapter, but the technique is especially well suited to dating stabilized sand dunes, which often mark old shorelines.

FACTORS AFFECTING LAKE LEVELS

The most important factors influencing lake level history are rebound of the Earth's crust, glacial margin fluctuations, drainage outlet location, and overall configuration of the basin. Each is explained below.

Crustal Rebound

Contrary to our own experience, the Earth's crust is not solid and unchanging but extremely sensitive to internal and external influences. For example, when Lake Mead was allowed to fill behind Hoover Dam in Nevada, the weight of the water caused the crust to subside about half an inch. If a large reservoir has this effect, imagine the consequences of a continental-sized glacier a mile thick. Scientists estimate that the Earth's crust was depressed nearly 1,000 feet in parts of northeastern Canada where the ice was thickest (Flint, 1971). As the ice melted, however, the weight was gradually reduced, and the crust began to rise, or "rebound," in response, much like the response of a foam chair cushion after a person gets up. The amount of rebound is directly related to the thickness of the ice at any particular spot—the thicker the ice, the greater the amount. In the case of the Laurentide Ice Sheet (the name given to the continental glacier that occupied northeastern North America), the ice was centered on the Hudson Bay region and became progressively thinner in all directions from its center. Thus, the amount of rebound that occurred at the glacier's periphery in, say, central Illinois, was much less than the amount around Hudson Bay. Accordingly, as one moves northward across the Great Lakes region, the amount of crustal rebound that has occurred progressively increases, reaching a maximum near Hudson Bay (Fig. 3.1). This effect is enormously important in understanding lake level history, as we shall soon see.

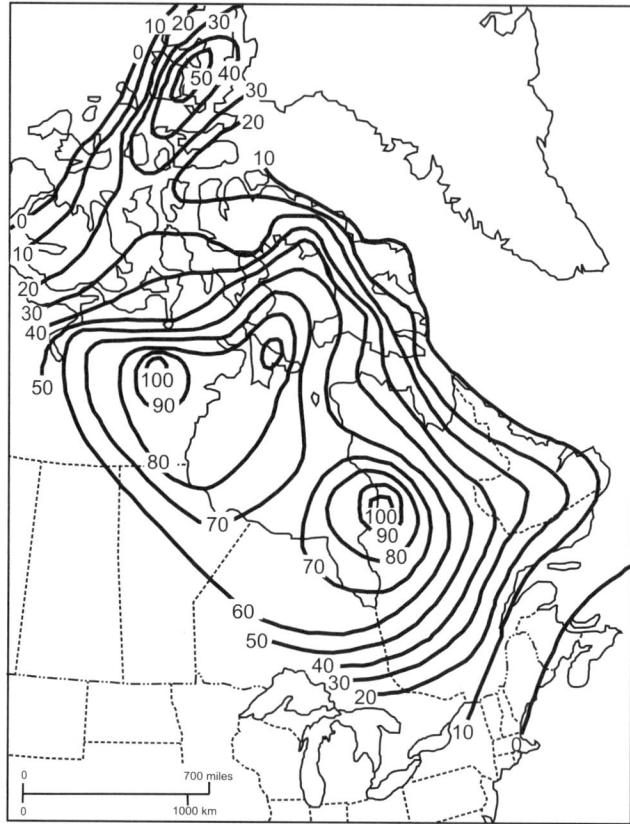

FIGURE 3.1. Crustal uplift (in meters) since the melting of the Wisconsin glacier (after Flint, 1971).

Ice-Marginal Retreat and Uncovering of Lower Outlets

Prior to glaciation, today's lake basins were probably broad river lowlands developed upon areas of weak rock (Spencer, 1891; Hough, 1958; Kincare and Larson, 2009). Glaciers advancing in broad tongues southward from Canada took the path of least resistance and followed these lowlands, grinding, plucking, and excavating deep troughs along the axes of the preexisting river valleys. Each time the glacier returned, it took a similar path, carving the basins ever deeper. By the beginning of the Wisconsinan glaciation, the last major ice advance into the region, it's likely that the rough configurations of today's Great Lakes basins were already in place. This situation is especially important to lake history because as the ice margin retreated northward across the region for the last time, meltwater collected in these deep depressions as they were uncovered by the ice, forming what geologists call proglacial lakes. These lakes expanded northward with the margin.

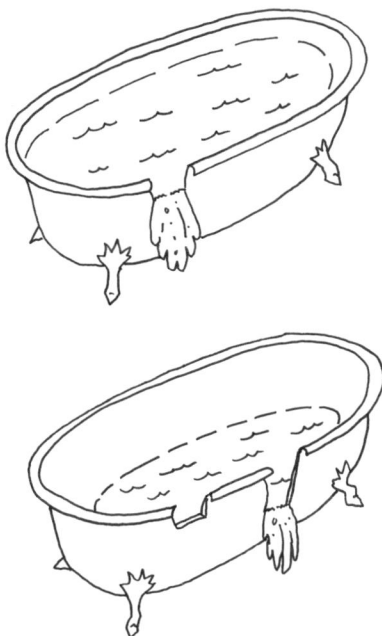

With continued melting, the level of the lake would rise until it was higher than the rim of land surrounding the basin, causing it to overflow at the lowest point along the rim. Geologists call these low spots outlets, and their elevation often controlled the level of the associated lake. The analogy is a bathtub in which a notch has been cut into the side (Fig. 3.2, top). The water would drain down to the level of the notch, but no farther. Even if water were slowly added to the tub from the faucet, the notch would continue to control the water level in the tub. If a deeper notch were cut (Fig. 3.2, bottom), the tub level would drain to

FIGURE 3.2. A notch cut in the side of a bathtub determines the water level in the tub. The lower the notch, the lower the water level. Each of the Great Lakes basins has its own set of notches, called "outlets" (from Blewett, 2009).

the new lower level, and so on. Like multiple notches in the side of a tub, each of the Great Lakes basins has its own set of outlets at varying elevations (Fig. 3.3).

When outlets undergo crustal rebound, the situation becomes much more interesting. We'll use Lake Michigan to illustrate the point. It so happens that the principal outlets of the Lake Michigan basin are found at either the northern or southern ends (Fig. 3.3). The southern outlet is called the Chicago outlet because of its proximity to that city. The present Straits of Mackinac mark the northernmost outlet. Imagine the situation as the ice margin retreated northward along the axis of the Lake Michigan basin (Fig. 3.4). In the south, a lake of ever-increasing size was formed in front of the ice as the glacier melted, spilling southward through the Chicago outlet (Figs. 3.4a, b). Eventually, however, the ice uncovered the Straits of Mackinac. This outlet was at a very low elevation, much lower then the Chicago outlet, because the crust in the Mackinac region had not yet rebounded, whereas the crust near Chicago already had begun to do so and was continuing to rise higher. Thus, the lake in the Lake Michigan basin spilled out through the Straits of Mackinac (a much lower notch in the tub), causing the water level in the basin to drop dramatically (Fig. 3.4c). With time, as the Mackinac area rebounded, this

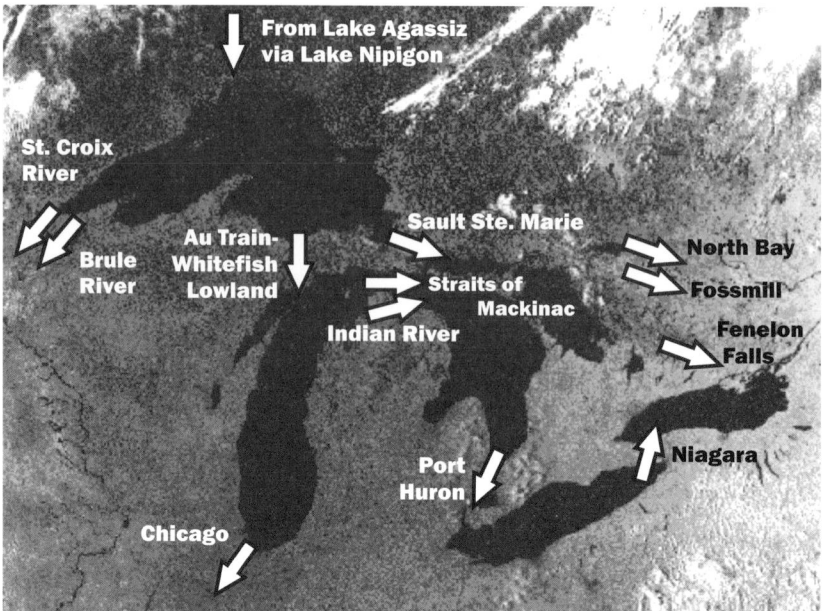

FIGURE 3.3. Principal outlets of the Great Lakes.

northern outlet began to rise, causing water levels in the Lake Michigan basin to rise. Water levels continued to rise until the waters began to slosh back southward toward Chicago, eventually reopening the southern outlet (Hansel and others, 1985).

Obviously, as with our example above, the shape and orientation of the major lake basins, along with the relative locations of their outlets, helped control the lake sequence. Figure 3.3 shows the major outlets for the Superior, Michigan, Huron, and Erie basins. Because water levels in the Lake Michigan and Huron basins were often confluent (as they are today), these two lakes typically were controlled by outlets east of Georgian Bay in Ontario (Fossmill, Fenelon Falls, North Bay outlets, Fig. 3.3). As these outlets rebounded, water sloshed southward toward Chicago or the present outlet at Port Huron. Details of the lake sequence are described in a later section.

Shoreline Patterns—Rebound and the Bathtub Revisited

How does rebound affect the pattern of ancient shorelines in a particular basin? The answer can be illustrated by placing our bathtub on a fulcrum, so that one side can be lifted up, mimicking the effects of rebound. We'll assume that our basin is oriented north–south, so that in figures 3.5a–c, north is to the right and south is to the left. Because rebound increases the farther north you go in this region, we will progressively "lift" the right side of the bathtub as time goes on to see what happens to the position of the shorelines over time. The actual situation is more complicated than this, but it serves to illustrate the point. Before rebound commences, the shorelines of our lake are found at the positions shown by line A. As the area rebounds, A rises in the north and sinks in the south, and a new lake is formed at level B (Fig. 3.5b). The process continues with lake C (Fig. 3.5c). Eventually, rebound slows and lake levels stabilize. Notice that shorelines A and B are preserved on the north side of the basin, whereas these same shorelines are submerged beneath the lake on the southern end in figure 3.5c. This situation is analogous to a coffee cup as we raise it to our lips. As we tip the cup, the coffee floods the side of the cup closest to us, while it falls along the opposite side. This simple model can be applied

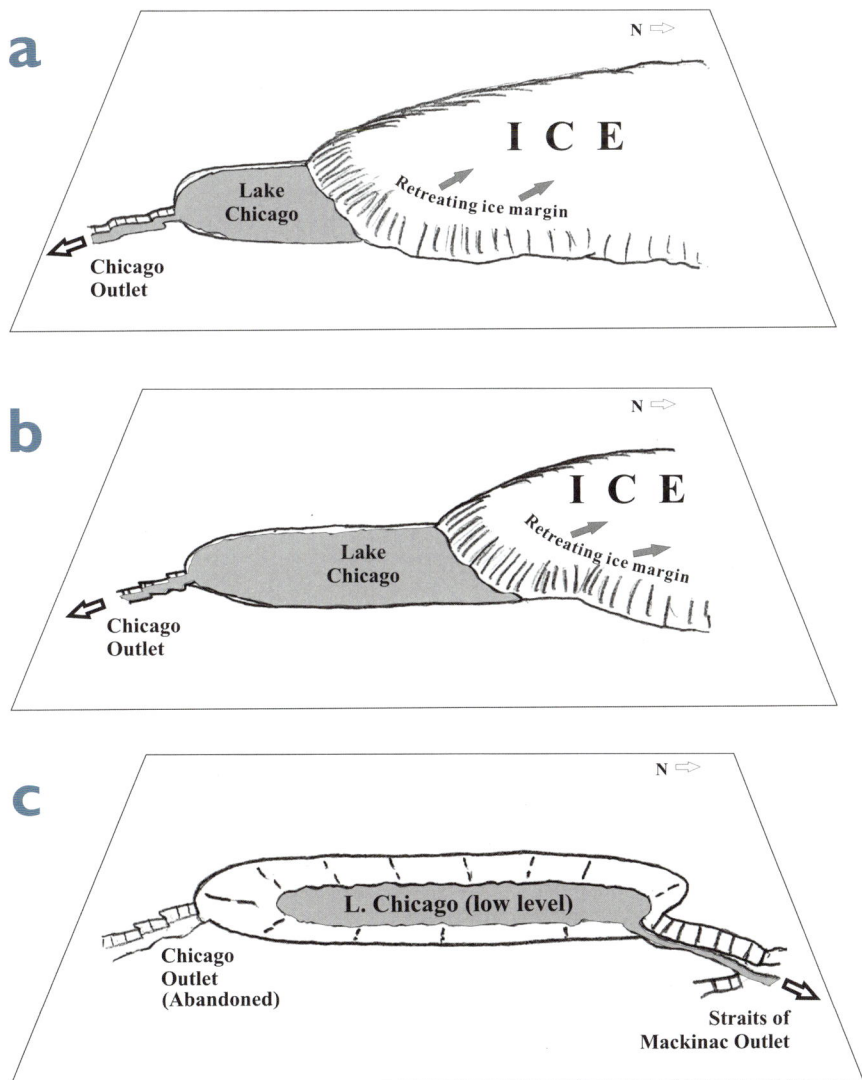

FIGURE 3.4 a, b, and c. With retreat of the glacial margin, the Straits of Mackinac outlet was uncovered, causing Lake Chicago to abandon the Chicago outlet and drop to a much lower level (from Blewett, 2009).

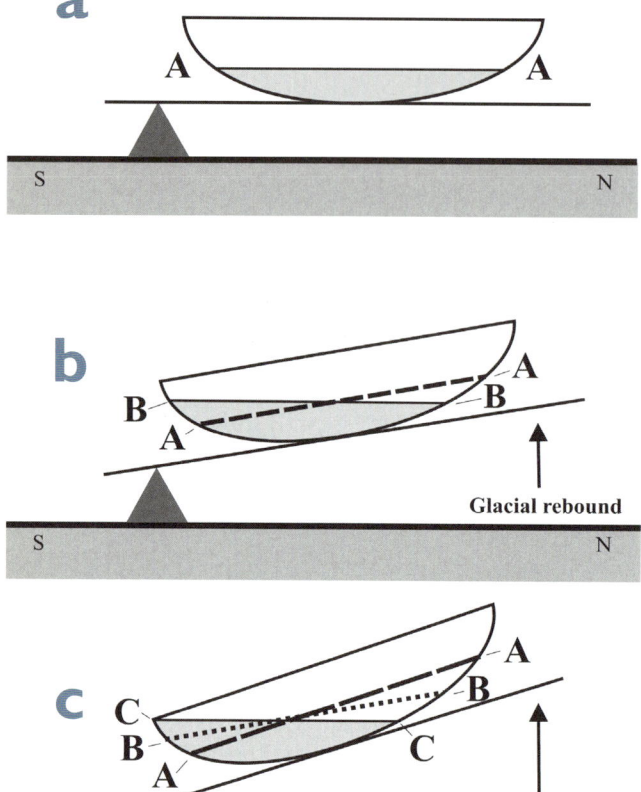

FIGURE 3.5 a, b, and c. Diagram illustrating the effect of rebound on the pattern of shorelines produced in a lake basin (after Larsen, 1987). The actual situation was much more complicated, with the fulcrum migrating northward across the basin.

to both the Lake Michigan and Lake Superior basins (with some important modifications) and explains why relict shorelines are the most numerous and well-developed along the northern coasts of Lakes Michigan and Superior, as at Pukaskwa National Park in Ontario. This model has its limitations, however, because in actuality the "fulcrum" as shown in figure 3.5 migrates northward with time.

Other Factors

Two additional factors affecting lake levels include changes in meltwater volume and the erosion of outlets. As glaciers melt, the volume of meltwater they release varies depending on temperature, precipitation, humidity, glacier surface area, and other factors. During periods of rapid melting, large volumes of meltwater would have been discharged into the adjacent basins, raising lake levels. Conversely, during times of slower melting, lake levels may have dropped. These changes could have occurred seasonally or over longer time periods. Because these adjustments to lake level were often short-lived, little time existed to produce well-developed shorelines, and geologists have been hard-pressed to verify this effect with any reliability. More certain is the effect that meltwater had on the erosion of outlets. Some, such as the one at Chicago, were initially cut into loose, easily eroded glacial sediments (gravel, sand, silt, and clay), which produced a lake at that particular level. As erosion of the outlet bottom proceeded, the lake began to drop until water eventually encountered the resistant bedrock lying beneath the glacial sediments, and the elevation of the lake stabilized (Hough, 1958).

All of these factors are responsible for the relict shorelines we see today, but they often combine in complicated ways that defy simple explanation. Accordingly, the history presented below is highly simplified, but it hopefully gives a general understanding of the main events recognized by glacial geologists.

THE LAKE SEQUENCE IN THE LAKE SUPERIOR BASIN

Events in the Lake Superior basin often operated independently of those in the Lake Michigan and Huron basins because lakes in the Superior basin drained through essentially one outlet, the rapids of the St. Marys River at Sault Ste. Marie, during most of their history. In addition, rebound increased from southwest to northeast across the basin, much like lakes farther south, but here the outlet at Sault Ste. Marie was located off at the basin's southeastern edge, making its influence on lake levels much more difficult to visualize. In effect, these relationships give the appearance that the basin "pivots" along a northwest–southeast line drawn from the Sault to Pigeon River on the Minnesota-Ontario boundary (Fig. 3.6) (Farrand and Drexler, 1985). Shorelines to the north and east of this line were raised progressively higher by rebound over time, while shorelines to the west and south of this line were submerged. The analogy is to a bathtub in which the drain is located along the side midway between the faucet and the foot of the tub. As one end of the tub is raised due to rebound, a familiar pattern of raised and submerged shorelines is developed on either end (Fig. 3.5), with the outlet located at the point where the shorelines converge.

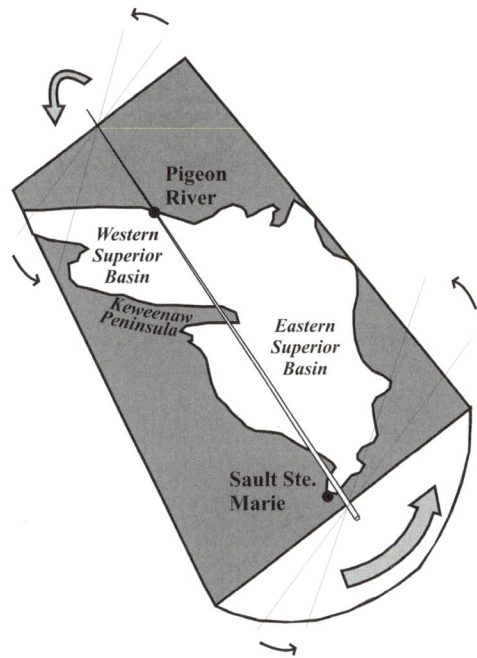

FIGURE 3.6. A simplified way to understand relations among Lake Superior basin shorelines, the outlet at Sault Ste. Marie, and glacial rebound. The basin appears to pivot on an axis drawn through Sault Ste. Marie, Michigan, and Pigeon River, Minnesota (from Blewett, 2009).

Recognized Shorelines

The Lake Superior basin can be divided into two parts: the western Superior basin, encompassing that part of the lake west of the Keweenaw Peninsula, and the eastern Superior basin, to the east of the Keweenaw (Fig. 3.6). As long as the ice margin remained south of the tip of the Keweenaw, lakes in the western and eastern parts of the basin remained separate (Figs. 3.8a, b). Today, a number of ancient shorelines exist high above the modern lake level in the western part of the basin. The name "glacial Lake Duluth" was proposed for the highest and most prominent of these

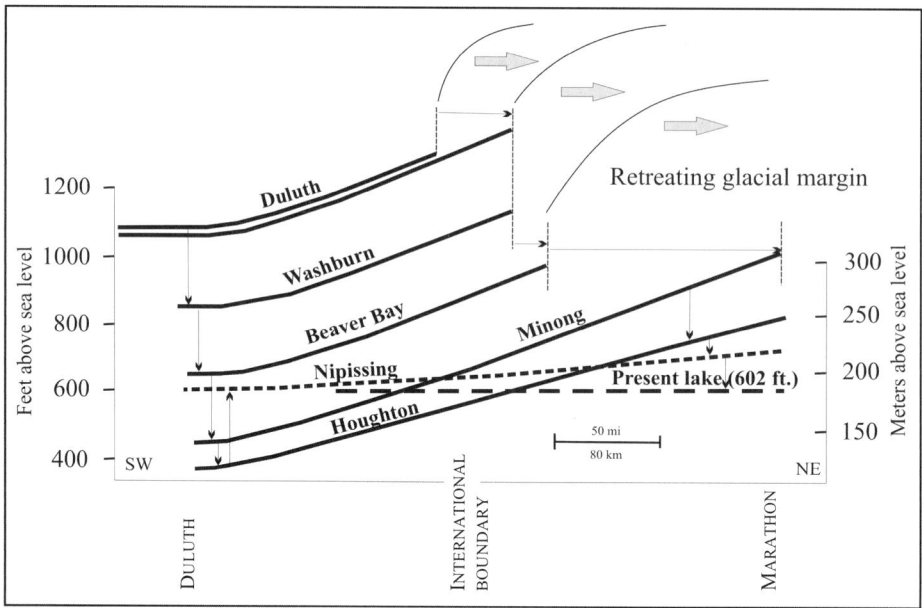

FIGURE 3.7. **Selected shorelines recognized in the Lake Superior basin (after Farrand and Drexler, 1985; Highbridge, Huron Mountain, and Algoma shorelines not shown).**

shorelines visible in the city of Duluth (Fig. 3.7). A slightly lower shoreline, likely associated with a minor retreat of the ice margin, is named "sub-Duluth." Below the Duluth beaches are a series of shorelines named Highbridge, Washburn, Beaver Bay, and Huron Mountain, each representing progressively lower water levels (Fig. 3.7) (Farrand and Drexler, 1985). In the eastern part of the basin a different set of shorelines exists; the shorelines are named the Minong, Houghton, Nipissing, and Algoma. As noted in figure 3.6, however, some of these shorelines are only visible northeast of the Sault Ste. Marie–Pigeon River line, the features having been covered by the modern lake southwest of this line. Figure 3.7 shows the relationship between the western and eastern basin shorelines. Now that a general idea of the existing strandlines has been presented, we'll turn to the lake chronology.

Sequence of Events

A number of glacial lakes likely occupied the Superior basin during final deglaciation, but an important glacial readvance, called the Marquette advance (discussed in the next chapter), obliterated most of their shorelines as it invaded the Superior basin approximately 11,500 years ago. This advance did not reach the Lake Michigan and Huron basins, but it likely filled the Lake Superior basin. Accordingly, our history begins with the first lakes to emerge as the Marquette ice front began to retreat from the southwestern edge of the basin just after 11,500 years ago.

Initially, a series of ice marginal lakes were dammed between the bedrock highlands to the south and the ice margin. These lakes spilled westward across the

axis of the Bayfield Peninsula and drained via the Brule outlet into the St. Croix River
(Figs. 3.3, 3.8a). Their shorelines are visible today in the vicinity of Apostle Islands
National Lakeshore. With continued retreat, these lakes coalesced to form glacial Lake
Duluth, which expanded northward with the wasting ice margin. This lake dropped to
a slightly lower level, the sub-Duluth, possibly as a result of downcutting of the Brule
outlet, and remained there until deglaciation of the Huron Mountains in Michigan
allowed the lake to find lower outlets to the east. The lake then rapidly drained through
a cascading series of lower lake levels, represented by the Highbridge, Washburn,
Beaver Bay, and Huron Mountain (Fig. 3.8b) shorelines (Farrand and Drexler, 1985).

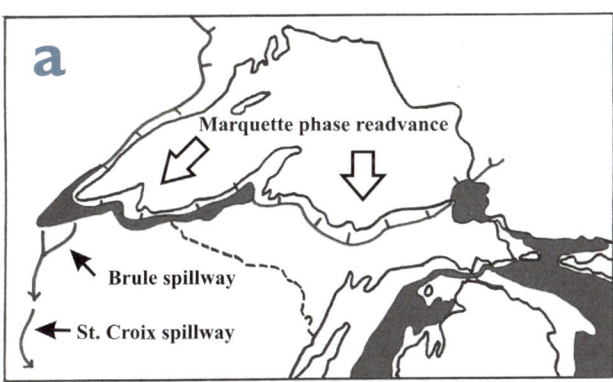

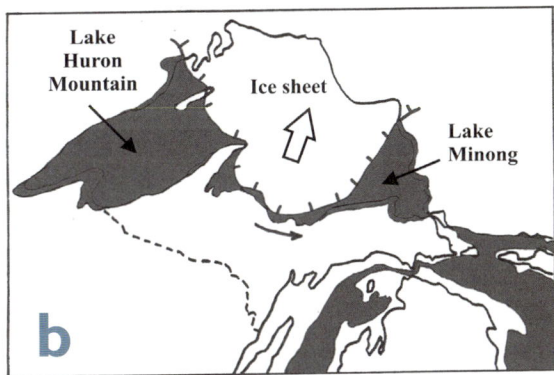

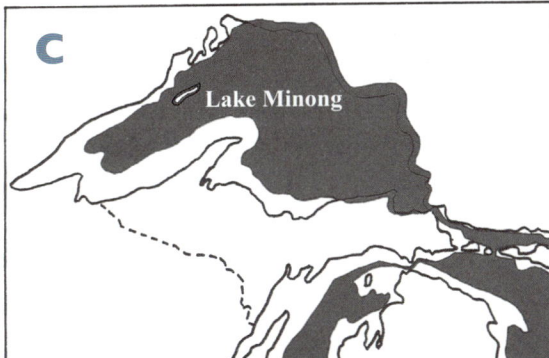

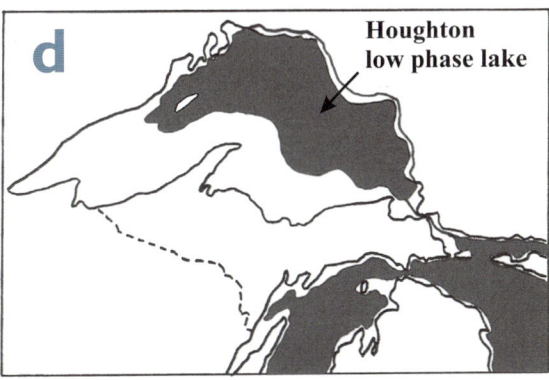

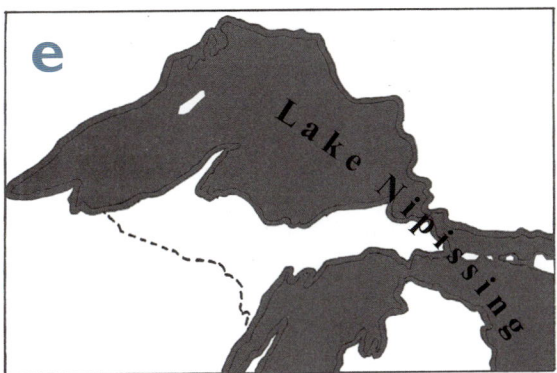

FIGURE 3.8. Simplified sequence of events in the Lake
Superior basin.

   a. Lake Duluth, 11,500 years ago at the peak of the
      Marquette ice advance.

   b. Lake Huron Mountain (a post-Duluth lake) and
      the beginnings of Lake Minong, about 11,000
      years ago. The Marquette ice margin is retreating
      into the Superior basin.

   c. Lake Minong approximately 10,700 years ago.

   d. Houghton low phase about 9,300 years ago.

   e. Lake Nipissing approximately 5,000 years ago.

Meanwhile, a different lake, named Lake Minong, occupied the now rapidly deglaciating southeastern end of the basin near Sault Ste. Marie by about 11,000 years ago. The ice margin still remained along the southern shore of Michigan's Upper Peninsula in the vicinity of Pictured Rocks National Lakeshore, however, and the lowest of the post-Duluth lakes drained along the edge of the ice through this area into an expanding Lake Minong (Figs. 3.8b, c). Retreat of the ice margin from the area of Pictured Rocks allowed the two lakes to merge and form a single expanded lake at the level of Lake Minong, with its outlet at Sault Ste. Marie. With final deglaciation, Lake Minong expanded across the entire basin, becoming the first post-glacial ancestor of Lake Superior (Fig. 3.8c). Due to rebound, Minong shorelines today are found northeastward of the Sault Ste. Marie–Pigeon River line mentioned previously. Southwest of this line they are submerged below modern lake level, except for the easternmost parts of Michigan's Upper Peninsula. Interestingly, shorelines related to Lake Minong, if projected southeastward toward the Sault, are 120 feet above the present outlet at Sault Ste. Marie. This means that the controlling outlet at Sault Ste. Marie must have been 120 feet above its present elevation at the time Lake Minong existed. To explain this difference, geologists have proposed that the Sault Ste. Marie area was likely covered by a thick accumulation of glacial material (called drift by geologists, and consisting of mixtures of loose boulders, sand, silt, and clay) that originally controlled the outlet of Lake Minong at the higher level (Farrand and Drexler, 1985). Until recently, geologists were at a loss to explain how this drift was removed to allow Lake Minong to drop to lower levels. Researchers working in Canada, however, have provided a possible answer.

It seems that about the same time that Lake Minong existed, a vast glacial lake, Lake Agassiz, covered parts of Saskatchewan, Manitoba, Ontario, Minnesota, North Dakota, and South Dakota. Modern Lake Winnipeg in Manitoba is a remnant of this lake. Detailed work on Lake Agassiz's shorelines, deposits, and outlet channels by various workers (Clayton, 1983; Teller and Thorleifson, 1983) indicates that Lake Agassiz drained into the Lake Superior basin, perhaps in huge floods, sometime during the Minong phase. If so, this flood of water from Lake Agassiz could have swept away the drift dam at the Sault, causing Lake Minong to drop to the level of the now denuded outlet. Whatever the cause, abundant evidence indicates that Lake Minong did drop over the succeeding several hundred years, eventually reaching a low-level lake called the Houghton low phase (Fig. 3.8d). This lake was controlled by the now eroded outlet at the Sault, which apparently was developed on bedrock after the higher drift dam had been swept away.

At the same time, the Nipissing transition was causing lakes in the Lake Michigan–Huron basin to rise higher and higher, partly due to the continued rebound of the controlling outlet in the North Bay region in Ontario (Larsen, 1985). Eventually these rising waters reached the level of Sault Ste. Marie, and Lake Nipissing expanded into the Lake Superior basin, erasing many of the Houghton low features and building new shorelines at higher levels. At this time the Sault would have been a strait, much like the present Straits of Mackinac, connecting Lakes Huron and Superior (Fig. 3.8e). Closely associated with the Nipissing, but lower in altitude, is another, fainter, shoreline named the Algoma, which records a drop in lake level from the Nipissing level approximately 3,400 years ago (Kincare and Larson, 2009). Until recently, geologists assumed that the level of Lake Nipissing was controlled by the Chicago and Port Huron outlets farther south, and that subsequent erosion of these outlets caused a drop to the level of Lake Algoma. More recent work indicates that changes in precipitation and climate, along with subtle adjustments of the outlet channels to rebound and changes in water volume, may be responsible for explaining the Nipissing and Algoma high lake levels (Fraser and others, 1990; Baedke and Thompson, 2000; Booth and others, 2002). In this sense, these Holocene water bodies behaved much more like the present Great Lakes, in which levels fluctuate based on rainfall and runoff generated by subtle annual variations in climate, although the Nipissing and Algoma fluctuations were much more pronounced. The Algoma level was common to all three basins (Superior, Lake Michigan, and Huron), but by approximately 1,500 to 2,280 years ago (Farrand and Drexler, 1985; Johnston and others, 2007) the rebounding outlet at Sault Ste. Marie raised water levels in the Lake Superior basin higher than those in the Michigan and Huron basins, creating modern Lake Superior. Initially, Lake Superior was slightly higher than present, at a level geologists call the Sault level. Climatic changes, erosion of the outlet at Sault Ste. Marie, or both may have led to a slight drop to the present level.

THE MODERN GREAT LAKES

Levels of the modern Great Lakes continue to fluctuate annually, but within a range much reduced from the mid-Holocene. Lake level gauges, emplaced in the 1850s, continue to measure changes in water levels and to provide an unbroken record of lake fluctuations spanning more than 150 years (Fig. 3.9a). They indicate that Lake Michigan levels typically range about 1 foot (high versus low) from the long-term average and have never exceeded more than 7 feet during the period of record. Geologists from the Indiana Geological Survey have been able to extend this record back in time by carefully measuring the altitude of old shorelines along the Lake Michigan basin and then dating the shoreline using radiocarbon techniques (Thompson and Baedke, 1995, 1997). These data are then graphed to produce a curve of former water levels (Fig. 3.9b). In analyzing the data, scientists were surprised to find that Lake Michigan undergoes a 160-year lake-level cycle in which levels broadly rise and then fall over this time frame. The data suggest that the relatively recent high lake levels of 1986 may have marked the crest of the cycle and that lake levels will continue to fall over the next 75 years. Data from beaches along Lake Superior show a similar pattern, although the fluctuation signal is weaker and determining the length of the cycle is much more difficult (Johnston and others, 2000; Johnston and others, 2002).

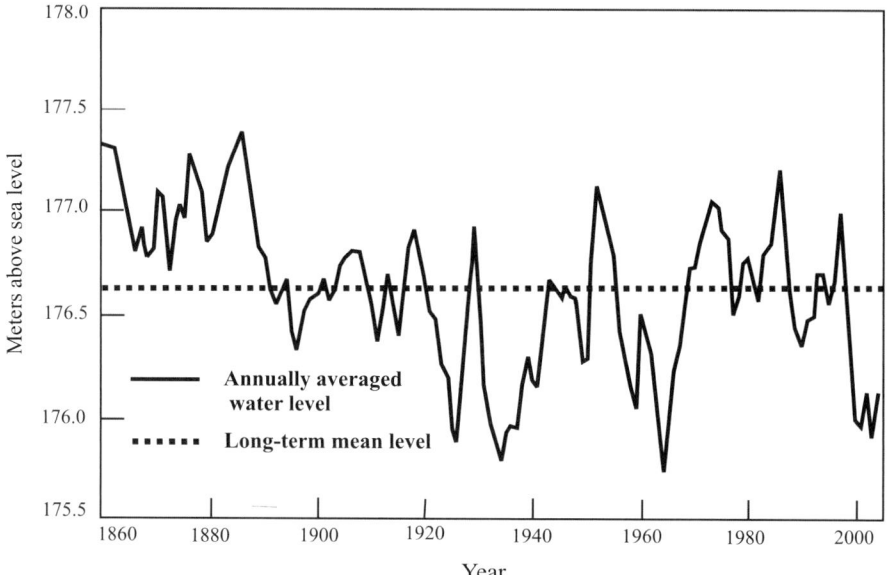

FIGURE 3.9a. Historic water levels of lakes Michigan and Huron based on lake level gauges.

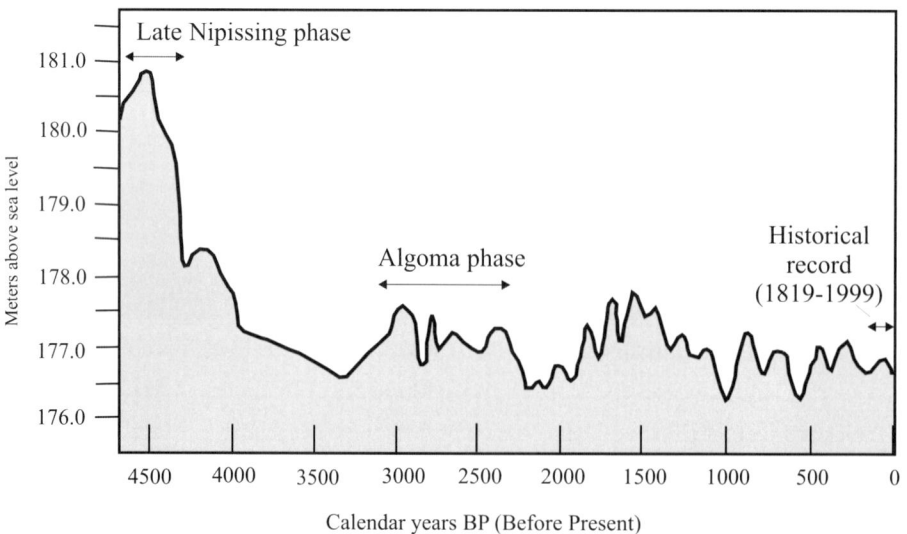

FIGURE 3.9b. Smoothed long-term lake levels in the Lake Michigan basin (after Baedke and Thompson, 2000).

FEATURES RELATED TO ANCIENT LAKE LEVELS IN PICTURED ROCKS NATIONAL LAKESHORE AND VICINITY

As mentioned previously, ice lingered in the area of Pictured Rocks long after the southwestern and southeastern parts of the Superior basin were ice free (Fig. 3.8b). By about 11,000 years ago, glacial lakes in the southwestern part of the basin drained along the southern edge of the ice, through the Pictured Rocks area, and into an expanding Lake Minong in the southeastern part of the basin (Farrand and Drexler, 1985). These drainageways (described in detail in chapter 4) were trapped between the ice margin and higher ground to the south (Hughes, 1968; Blewett, 1994). They also carried enormous amounts of sand and gravel, which were dumped along their

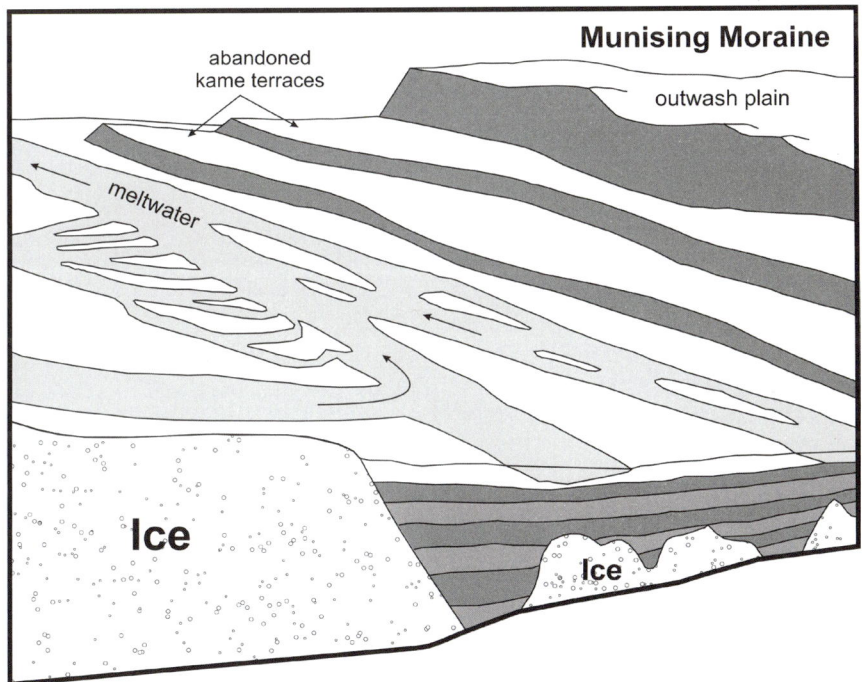

FIGURE 3.10. Diagram showing the formation of meltwater terraces near Grand Marais in Pictured Rocks National Lakeshore (from a drawing by Gregg Bruff). View is southeastward.

bottoms. As the ice continued to retreat into the Superior basin, it uncovered a series of progressively lower outlets, and new, lower drainageways formed. This situation resulted in a stair-step-like series of sand- and gravel-floored terraces (called kame terraces), recording successive drops to lower levels (Fig. 3.10). These features are not shorelines, but they do record the sequence of drainageways that linked the various lakes along the southern margin of Marquette ice during its retreat.

These terraces are best viewed at the Sable Falls parking lot near Grand Marais. The parking lot is located on a 720-foot terrace. The hill behind the parking lot is a "riser" up to a 750-foot terrace (Figs. 3.11, 3.12a, b; see road log). The terraces can also be viewed by driving south on County Road H-58 from the Sable Falls parking

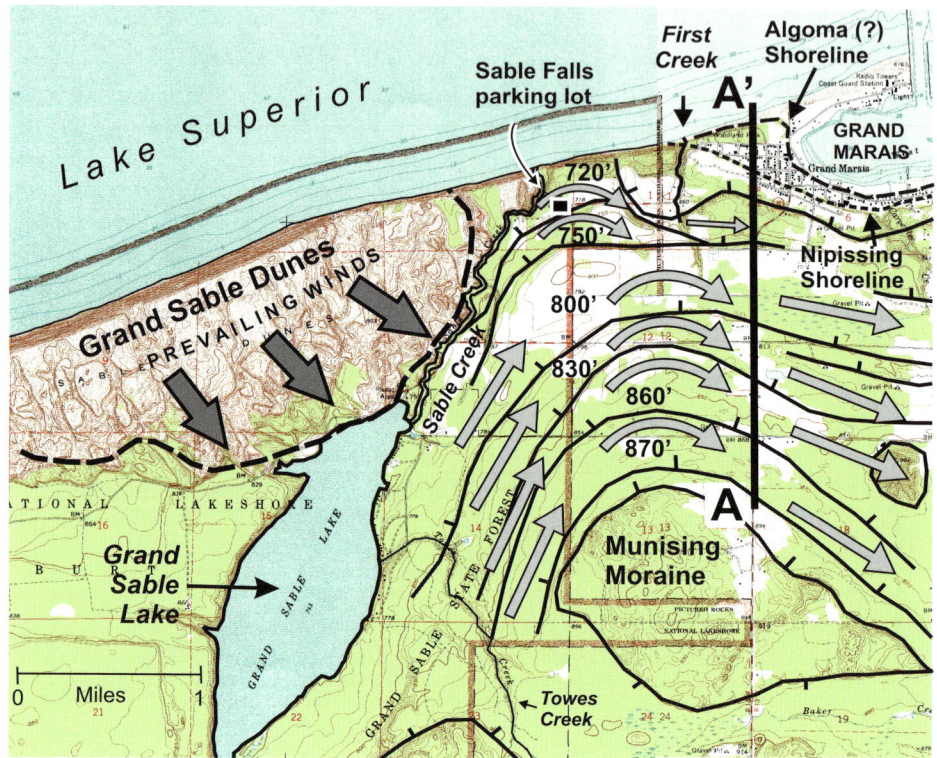

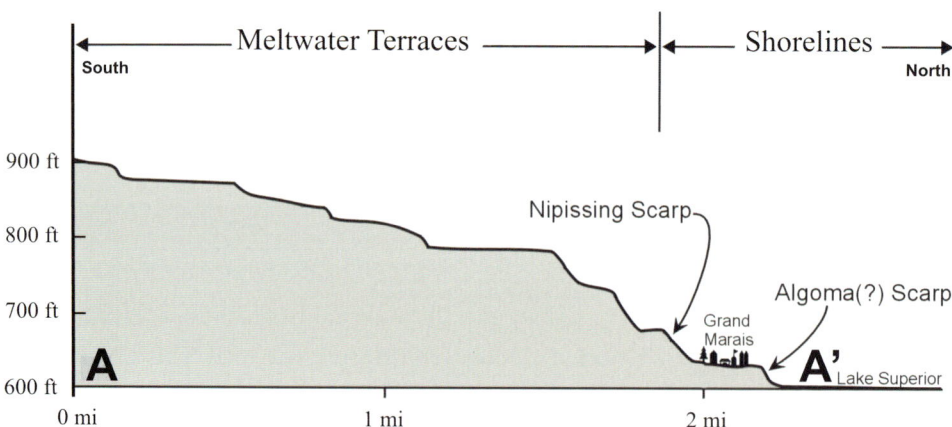

FIGURE 3.11. Meltwater terraces and landform relationships in the vicinity of Grand Sable Lake. Line A-A' (top image) marks the location of the profile shown in the bottom diagram (after Blewett, 2002b).

lot. Early scientists, working without the benefit of topographic maps, misinterpreted these terrace scarps as ancient shorelines. More recent work suggests that the only true ancient shorelines in the park are associated with the Nipissing and Algoma phases, although W. L. Loope (personal communication, 2010) indicates that traces of the Minong shoreline may be present.

The Nipissing shoreline is well defined throughout the park and is often seen as a conspicuous "notch" in the shoreline profile about 40 feet above the present lake level (Fig. 3.13). This lake had a number of small embayments that are defined by the bluff faces south of the present shoreline along Miners Beach, Chapel Beach, and Beaver and Trappers Lakes. Depositional features such as beach ridges and bars are also common. The alcoves of Chapel Rock were likely initiated as shoreline caves cut by the waves of Lake Nipissing (Dorr and Eschman, 1970), although differential weathering and erosion have also played a role in shaping the landform. Miners Castle is often erroneously identified as an ancient shore feature. Its configuration is more likely due to weathering and erosion and the peculiarities of the bedrock from which it is made (Anderton, 2002) (see chapter 5). Beach ridges related to Lake Nipissing and the post-Nipissing transition are also well displayed in the region, often as a series of corrugated parallel ridges located between the Nipissing beach and the modern lake. They are most easily observed along the trail from Little Beaver Lake campground to Lake Superior, near Au Sable Point, at Chapel Beach, and along the Sand Point Marsh Trail.

FIGURE 3.12a. Kame terraces on county road H-58 at the turnoff for Sable Falls. Photo is looking south from the 720-foot terrace. Cars are on the 750-foot terrace, and the top of the road in the background is on the 800-foot terrace (see Fig. 3.11). Drainage was to the left (east).

FIGURE 3.12b. Kame terraces at the Sable Falls parking area. Flat area in the foreground is the 720-foot terrace. The bluff in the background is the riser to the next highest terrace at 750 feet. Drainage was from right to left (east). Note person for scale (arrow).

FIGURE 3.13. Nipissing shoreline "notches" (arrows) in the Au Sable Point headlands, as seen from the Sable Falls beach.

Nipissing and Algoma features are well displayed in both Grand Marais and Munising (Blewett, 2002b). In Grand Marais, the steep bluff immediately south of downtown is the Nipissing storm beach (Figs. 3.11, 3.14). The main part of the village sits on old Nipissing lake bottom. The road from downtown to Coast Guard Point drops down an Algoma shoreline scarp just as it leaves the main part of downtown. The city marina sits at the base of this scarp on old Algoma lake bottom (Fig. 3.15) (see road log). This scarp can be followed into the eastern part of Grand Marais and is well displayed where steps climb the entrance to the Burt Township school (Fig. 3.16). Likewise, most of the city of Munising is built upon old Nipissing lake bottom. State Highway M-28 skirts the north side of the business district on a spit or barrier associated with this or a slightly lower lake level (Fig. 3.17). Several distinct terraces associated with the Algoma (?) or slightly lower levels can be observed in Bayshore Park across from the Pictured Rocks Boat Tours (Fig. 3.18). Grand Island, of course, displays one of the finest tombolos in the region. A tombolo is an isthmus that connects an island with the mainland, or, as in the case of the Grand Island tombolo, to another island (Figure 3.19). The tombolo on Grand Island is another excellent place to observe corrugated beach ridges marking the post-Nipissing transition.

FIGURE 3.14. Downtown Grand Marais is built on the former lake bottom of Lake Nipissing. The Nipissing shoreline is located at the base of the hill in the background.

FIGURE 3.15. Post-Nipissing beach scarp near the Grand Marais marina, possibly related to the Algoma lake phase.

FIGURE 3.16. Stairs climb the post-Nipissing beach scarp at the entrance to the Burt Township School just east of downtown Grand Marais.

FIGURE 3.17. A Nipissing or post-Nipissing spit or barrier at the northern end of the Munising business district. View is northward (toward the bay) along Elm Street.

FIGURE 3.18. Park benches sit on a post-Nipissing (Algoma?) beach scarp at Bayshore Park in Munising, across from the Pictured Rocks Boat Tours dock.

FIGURE 3.19. Oblique aerial view of Grand Island with its famous tombolo. The east arm of Munising Bay is in the foreground. (Courtesy of National Park Service)

4

THE ICE AGE

With the coming of the Great Ice Age, the last chapters in the geologic history of Pictured Rocks National Lakeshore begin. Glaciers advancing southward from source regions in eastern Canada destroyed the preexisting landscape as they excavated the basins of the Great Lakes, bulldozing away any remaining Paleozoic or Mesozoic rocks located above the Au Train Formation and mantling the region with thick accumulations of glacial drift. Each succeeding glaciation tended to take similar paths southward, reexcavating and deepening the Great Lakes basins with each advance, eventually producing the enormous depressions observed today. The effect was to essentially erase all that had come before, replacing it with a much lower relief landscape set amid the vast and exceedingly deep depressions of the newly created Great Lakes.

THE GLACIAL CHRONOLOGY

As we reviewed in chapter 1, the Ice Age only dates back a few million years and is part of the time period known as the Quaternary (qua-TER-na-ry) Period, which consists of two epochs, the Pleistocene and Holocene (see Fig. 1.6). Geologists generally define the Pleistocene as the time of the Great Ice Age and the Holocene as the present warm period that began when the glaciers melted. Originally, they envisioned four major glacial episodes within the Pleistocene that were named (from oldest to youngest) the Nebraskan, Kansan, Illinoian, and Wisconsinan glaciations. Recent discoveries indicate that the Ice Age actually began about 2.6 million years ago, possibly as a result of natural perturbations in the Earth's orbit that changed the relative amounts of solar radiation received during particular seasons. These orbitally induced climatic changes resulted in cycles of warming and cooling that produced a glaciation about once every 100,000 years over the succeeding 2.6 million years, or about 20 glaciations in all (Shackelton and Hall, 1984). Accordingly, the old idea of four glaciations was discarded, although the names Wisconsinan for the latest glaciation and Illinoian for the previous one have been retained. Glaciations older than the Illinoian are referred to as pre-Illinoian.

By glaciation we mean an event during which the ice accumulated in northeastern Canada, advanced southward into the present Great Lakes region, and then melted away. Because each succeeding glaciation erased much of the evidence of the preceding one, only the latest advance, the Wisconsin, is of importance to our discussion.

The Wisconsin glaciation began approximately 79,000 years ago and reached a maximum position approximating the latitude of the Ohio and Missouri Rivers about 23,900 years ago (Larson and Kincare, 2009). After this time, the ice margin began to retreat northward in response to a warming climate, punctuated by small readvances. By about 17,000 years ago, the ice margin had retreated to the southern edge of the Lake Michigan basin. By 13,600 years ago it was located near Two Rivers,

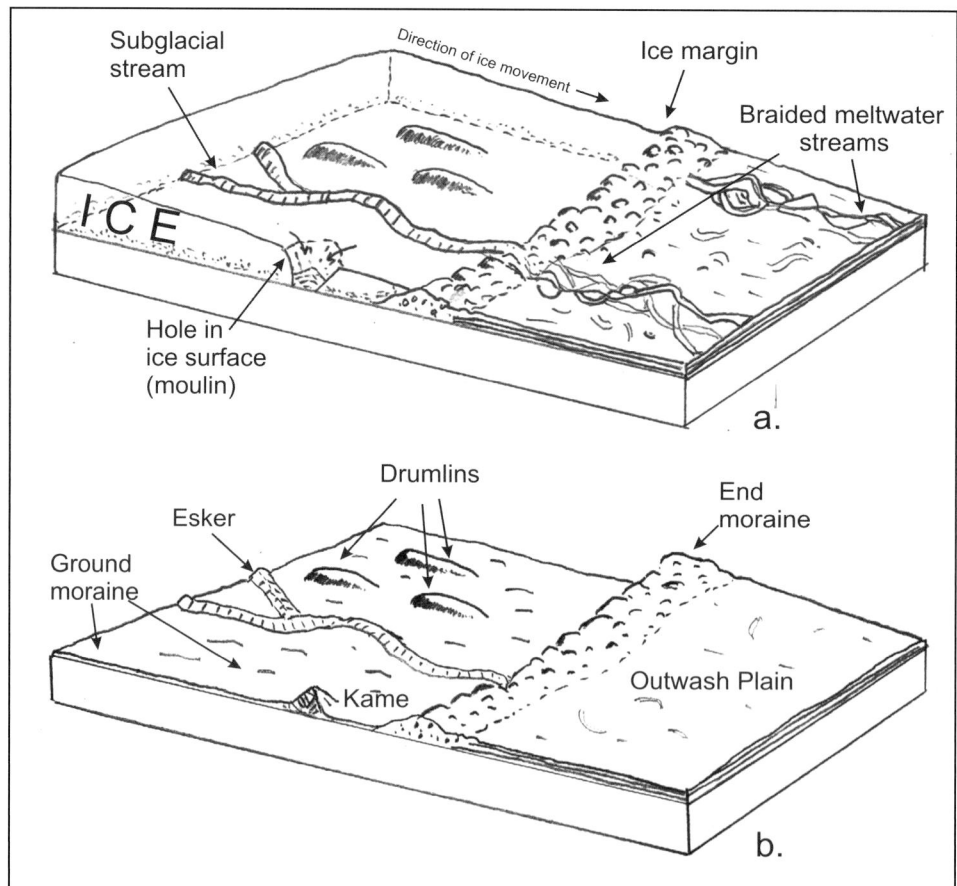

FIGURE 4.1. Formation of common glacial landforms found in Pictured Rocks National Lakeshore. Drawing by the author.

Wisconsin, and by 13,000 years it had probably reached the central Upper Peninsula (Larson and Kincare, 2009; Futyma, 1981).

For the record, glaciers never "retreat," only their margins do. As long as they are not stagnant (wasting away in place), glaciers are always moving forward under their own weight (much like pancake batter when first poured on a griddle). When the rate of melting at the margin exceeds the rate of forward movement, the margin retreats. If forward movement increases or melting slows, the glacier margin advances.

A GLACIER PRIMER

To understand the glacial history of Pictured Rocks, we first need to review some of the basic terminology associated with glacial geology. In a simple sense, glaciers act like giant bulldozers, scraping and transporting eroded rock material from beneath the ice and dumping it at their margins. If the rate of glacial advance equals the rate of melting, the edge of the ice may remain at a particular position for a prolonged period, resulting in the formation of a large pile of debris along the glacial margin. If the glacier then melts away, a steep linear hill called an end moraine, approximating the former ice marginal position, is left behind (Fig. 4.1). With continued retreat of the ice margin, a series of end moraines may be constructed. End moraines are typically made of a particular kind of drift called till (Fig. 4.2), defined as a mixture of angular-shaped gravel, sand, and clay deposited directly by the ice. Drumlins are

FIGURE 4.2. Red till, likely associated with the Marquette glacial advance approximately 11,500 years ago, exposed along Sable Creek.

cigar-shaped hills oriented parallel to glacial movement that form along the bottom of advancing ice. They are usually composed of till and are important indicators of glacier flow directions. Drumlins are often associated with extensive tracts of streamlined low hills and grooves oriented parallel to ice movement called flutes. Both drumlins and flutes are common in western sections of Pictured Rocks National Lakeshore (see Fig. 1.2a).

When the ice margin thins or becomes detached from the main ice sheet, it may cease forward movement and stagnate. In these situations, meltwater often collects along the bottom of the ice, forming streams in tunnels along the glacier bottom. These tunnels become choked with sand and gravel and, upon ablation (melting)

FIGURE 4.3. Outwash deposited by meltwater exposed in a gravel pit near Kalkaska, Michigan, in the northern Lower Peninsula. Note dominance of sand and gravel and distinct layering within the deposit.

of the glacier, are often left behind as sinuous hills called eskers (Fig. 4.1). Because eskers are deposited by meltwater rather than by the ice itself, their sediments have characteristics very different from till. The sediments are well sorted, rounded, stratified, dominated by sand and gravel, and called glaciofluvial sediment or outwash (Fig. 4.3). Melting glaciers also exhibit large holes or depressions at the surface, which may fill with outwash. After the ice melts away, the outwash is no longer supported by the ice and left behind as a small, steep, cone-shaped landform called a kame (Fig. 4.1). Both eskers and kames are common in the Pictured Rocks area.

Outwash plains form as sediment-choked meltwater streams deposit sand and gravel at and beyond the ice margin (Fig. 4.1). Where the area beyond the ice margin is not blocked by higher ground, the streams are free to deposit outwash across extensive tracts, forming broad outwash plains, such as the Kingston Plains, that slope gently away from the ice margin. Where the glacier margin is retreating down a steep slope, meltwater released from the glacier is often ponded between the margin and the adjacent slope and forced to flow along (or parallel) to the glacier's edge (see Fig. 3.10). Upon ablation, the accumulating outwash forms a narrow outwash plain oriented parallel to the former ice margin called a kame terrace (see Fig. 3.10). With continued retreat of the glacier margin down the slope, a series of kame terraces, each one lower than the one that preceded it, are left behind in stair-step-like fashion. As reviewed in the previous chapter, these landforms are common in Pictured Rocks National Lakeshore and form the majority of terrain in the eastern two-thirds of the park.

Closely related features called heads of outwash (Blewett and Rieck, 1987; Blewett and others, 2009) are also common in the Pictured Rocks region. Although their genesis is still a matter of debate, heads of outwash are perhaps best explained

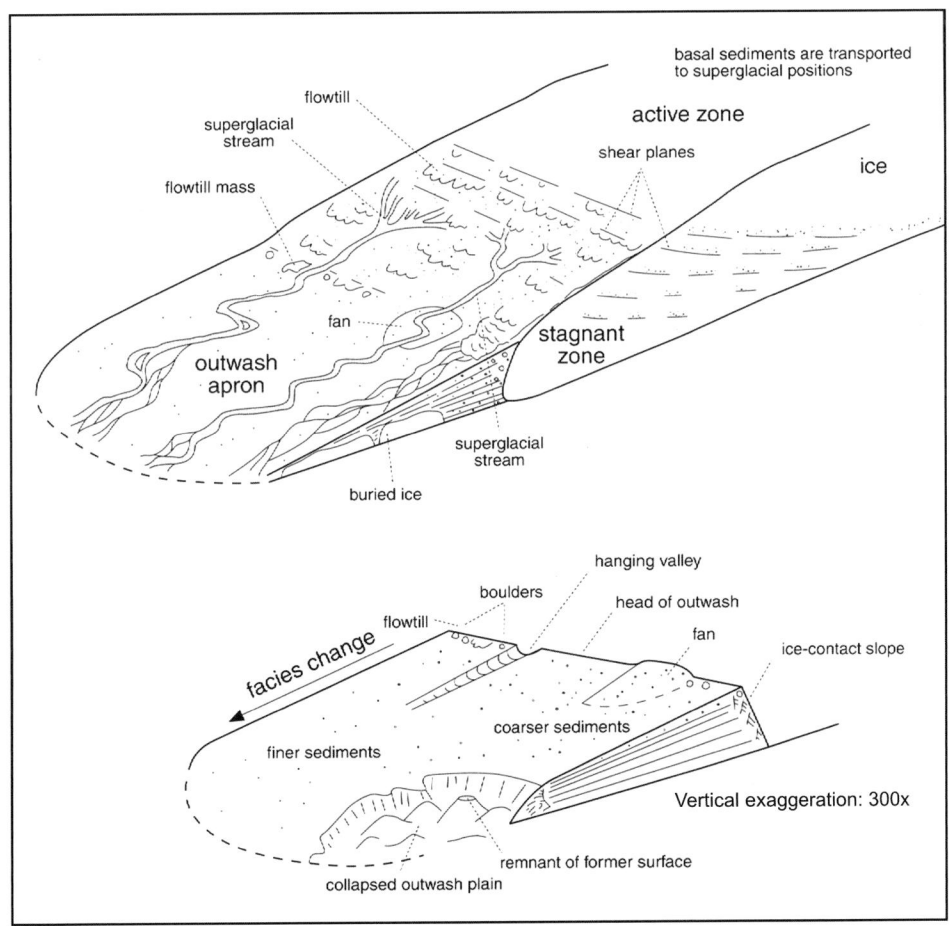

FIGURE 4.4. Genesis of heads of outwash as inferred from Koteff and Pessl (1981) (from Blewett and Winters, 1995). Many areas mapped as moraine in the vicinity of Pictured Rocks are better interpreted as heads of outwash and related ice wastage features.

using the stagnation-zone retreat model of deglaciation (Koteff and Pessl, 1981), in which a thin belt of stagnant ("dead") ice forms along the glacier's margin. Shear planes develop between the active ice and the stagnant marginal zone, allowing sediments from the glacier's base to be transported and dumped on top of ice in the stagnant zone (Fig. 4.4). Meltwater and mass movements then transport these sediments beyond the ice margin forming outwash aprons. Sediments associated with a head of outwash often change from coarse to fine textures as distance from the ice margin increases, with boulders deposited near the glacier's edge.

The area along the ice margin serves as the apron's head of outwash. Upon ablation, the outwash deposits, often mixed with small amounts of till, collapse to the angle of repose forming a steep hill called an ice-contact slope. The result is an asymmetrically shaped landform in profile, steepest on the up-ice side. As the border of active ice retreats, a series of heads of outwash may be left in the landscape, delimiting former ice-marginal positions (Fig. 4.5). Kettles (bowl-shaped

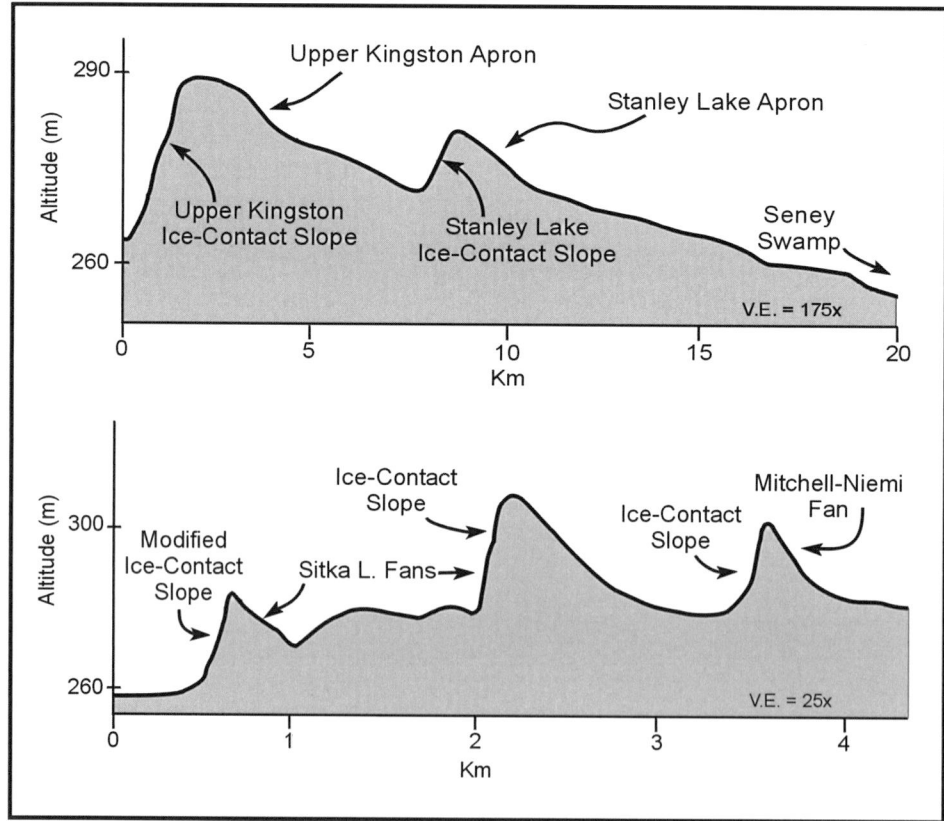

FIGURE 4.5. Profiles of distinct heads of outwash located within the Munising moraine just south of Pictured Rocks National Lakeshore. Each head of outwash represents a separate ice-marginal position during final retreat of the ice margin across the region (from Blewett and Rieck, 1987). Landform names are shown on Fig. 4.7.

depressions) eventually form by the melting of ice blocks buried beneath drift. Crevasse fillings (see below) and related features also form in the stagnant zone. Heads of outwash are distinguished from simple outwash plains by the complexity of sedimentation within the ice-contact zone, their mode of formation, and their tendency to be controlled by a particular base level.

Glaciers also develop large surficial fissures called crevasses, which may fill with glaciofluvial sediment washed in from the glacier's surface. When the glacier melts, these deposits are unsupported and collapse to form linear hills of glaciofluvial sediment called crevasse fillings.

Kettles are typically formed in association with a wasting glacier and are widespread in the Pictured Rocks area. Kettle chains are rows of kettles arranged in linear fashion (like pearls on a necklace) that are often found on outwash plains. The Kingston Lake kettle chain just south of the park is an excellent example of this type of landform (see below).

DRIFT AND BEDROCK SURFACE CHARACTERISTICS

With the exception of a narrow zone along Lake Superior, most of the Pictured Rocks area is mantled by glacial drift of varying thickness. R. L. Rieck (1991) compiled a map of the bedrock surface configuration of Alger County based on an analysis of waterwell and petroleum well logs. Once this map was compiled, he could then subtract the elevation of the bedrock surface from the modern topographic surface to determine drift thickness. The descriptions that follow are offered for those readers with a need for more detailed information on the drift characteristics.

Bedrock Topography

Inland between Munising and Grand Portal Point, a thinly mantled low-relief bedrock plateau at approximately 850 to 900 feet above sea level drops steeply lakeward to 328 feet at Sand Point. This plateau likely marks the dip slope of the subcropping Cambro-Ordovician escarpment. Farther east the plateau is interrupted by the 195-foot-deep north–south trending Miners River bedrock valley. A buried bedrock valley of similar size and trend may also exist near Kingston Lake (Blewett and Rieck, 1987). Near the park's eastern boundary, bedrock altitudes drop from 875 feet about 3 miles south of Grand Marais to less than 490 feet (112 feet below the level of Lake Superior) at the lakeshore.

Drift Thickness

Drift thickness increases from west to east in the park. South and west of Grand Portal Point, drift is absent or very thin within several miles of the shoreline, becoming thicker inland, but seldom exceeding 30 feet. Drift thickens eastward, reaching at least 80 feet along the lakeshore southwest of Twelvemile Beach Campground, about 100 feet southeast of the campground, 65 feet near the mouth of Sullivan Creek, and 80 feet along Hurricane River. Farther inland drift thickness likely exceeds 65 feet. South of Grand Marais, however, a broad bedrock highland is thinly mantled by drift, and in places bedrock crops out.

Drift Characteristics

Surficial sediment characteristics differ markedly between eastern and western sections of the park. Sandy till prevails inland south and west of Grand Portal Point (Hughes, 1968), whereas outwash predominates to the east, with very limited amounts of till. Glaciofluvial deposits are exposed at a gravel pit south of Beaver Lake, where rounded boulders up to 1 foot in diameter exist within a matrix of stratified sand and gravel. Areas mapped as coarse-textured till southeast of Kingston Lake (Farrand and Bell, 1982) are almost certainly glaciofluvial and related deposits (Blewett, 1994). A distinctive red till attributed to the Marquette readvance is exposed along the bank of Sable Creek near Sable Falls (Fig. 4.2; see road log for location).

LANDFORMS
Moraines, Outwash Plains, and Heads of Outwash

The Munising moraine (Leverett, 1929), a conspicuous east–west trending highland, skirts the southern edge of Pictured Rocks National Lakeshore (Fig. 4.6; see also Fig. 1.2b). Blewett and Rieck (1987) proposed that parts of this feature are better

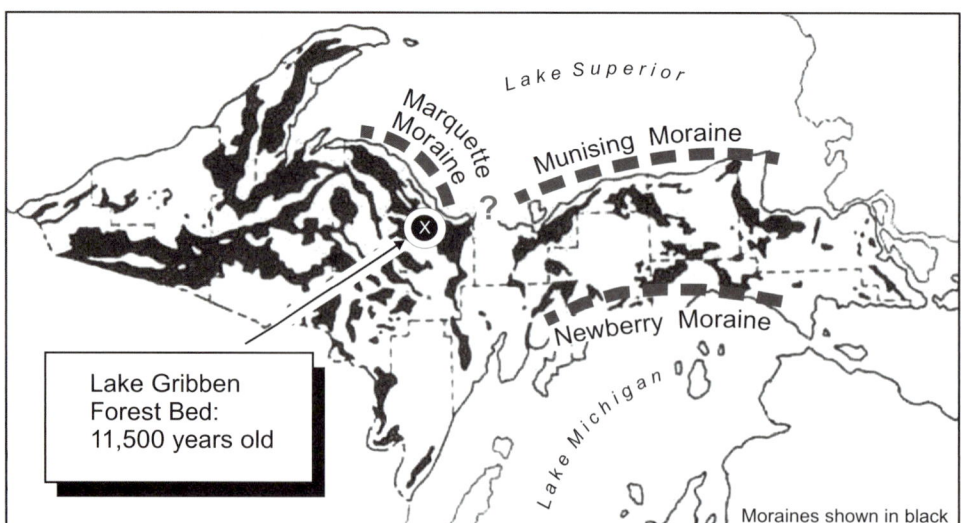

FIGURE 4.6. The two principal moraines of the eastern Upper Peninsula and their relation to the Lake Gribben forest site.

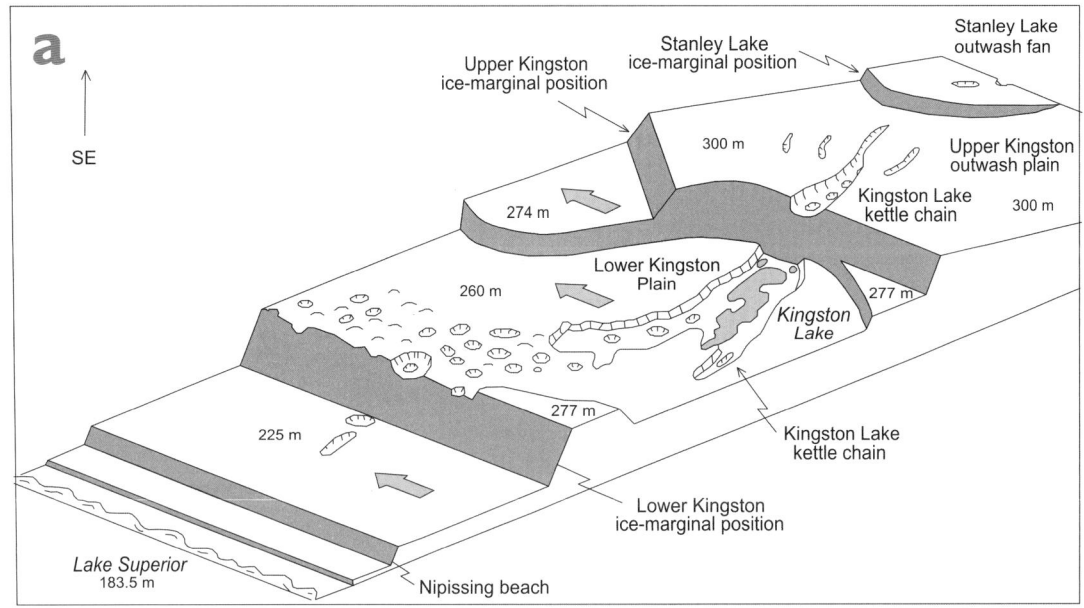

FIGURE 4.7 a, b. Two views of the Kingston Plains area. Figure 4.7a shows the principal landforms looking southeast (from Blewett and Rieck, 1987). Figure 4.7b shows the same general area looking northeastward (author's unpublished drawing, 1987).

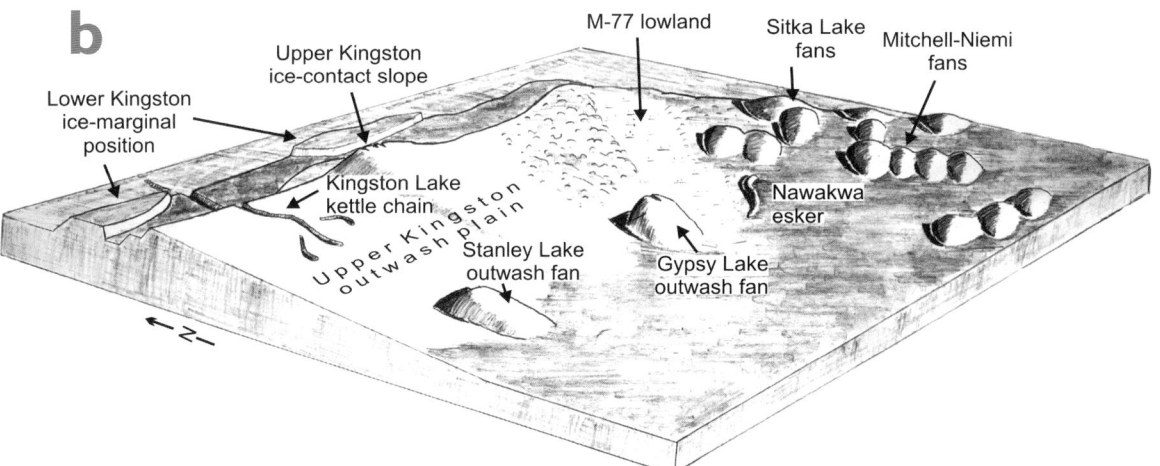

interpreted as heads of outwash rather than a moraine. Important heads of outwash in the Pictured Rocks area include the Stanley Lake, Upper Kingston, and Lower Kingston ice-marginal positions (Figs. 4.5, 4.7a, b). Most of the associated ice-contact slopes have been trimmed and significantly modified by meltwater streams. Only a few unmodified examples exist and most are not easily accessible, the Log Slide position being an important exception (see road log). Collectively, their widespread occurrence affirms the importance of ice-marginal stagnation and meltwater deposition in developing the region's glacial landscape.

Other ice wastage features are common south of the park border, and especially well-developed northeast of Melstrand, where a 2-mile-long, 50-foot-high esker grades into a large outwash fan (Figs. 4.8b, c; see also Fig. 1.2a). A related but more poorly developed esker is located nearby and is easily accessible from the road to the Chapel area trailhead (see road log). The Nawakawa esker is another especially well-developed esker located just east of M-77 between Grand Marais and Seney (Figs. 4.7b, 4.8a; see also Fig. 1.2b).

Kingston Lake Kettle Chain

One of the most significant and interesting glacial features in the region is the Kingston Lake kettle chain, which consists of a series of kettles forming the deeper parts of a nearly continuous depression 6 miles in length. The feature is oriented northwest–southeast and cuts across two distinct ice-marginal positions (Figs. 4.7, 4.8f). Depth of the dry kettles exceeds 50 feet and their bottom altitudes become progressively lower to the north. A small water body, informally named Turtle Lake (see road log), is the deepest kettle within the chain, exceeding 130 feet in depth. Except for Kingston Lake, kettles in the south are generally water-filled, while those to the north are dry. Near the middle of the Upper Kingston apron a branching occurs in the chain, somewhat resembling a dendritic stream pattern, with tributaries converging to the north (Figs. 4.7, 4.8f; see also Fig. 1.2b). Nugent, Ewatt, and Birch Lakes are water bodies in this secondary chain.

How does such a peculiar landscape form? W. L. Blewett and R. L. Rieck (1987), following a related mechanism proposed by Hughes (1968), interpreted the Kingston Lake kettle chain as the result of a stream valley cut into the surface that existed prior to the latest glaciation. As the retreating margin of the melting glacier reached this area, stagnant ice lingered in these valleys and was covered by outwash. Later, after the main ice sheet had withdrawn into the Lake Superior basin, the buried ice melted, leaving the linear kettles (Fig. 4.9). This preglacial valley, though buried, may have resembled the modern valley of Miners River, which is flowing in bedrock several miles to the west. Both features show similarities in depth, morphology, and trend of the valley bottom elevations. The Kingston Lake kettle chain is readily accessible from H-58, and few trees obscure the kettle morphology (see Fig. A.13 in the road log).

Drumlins

J. D. Hughes (1968) noted that extensive tracts between Melstrand and Munising previously mapped as moraine are actually northwest–southeast oriented flutes and drumlins (Figs. 4.8a–f; see Fig. 1.2a). No well-developed examples are visible or easily accessible from the park roads, but the general rolling character of the topography

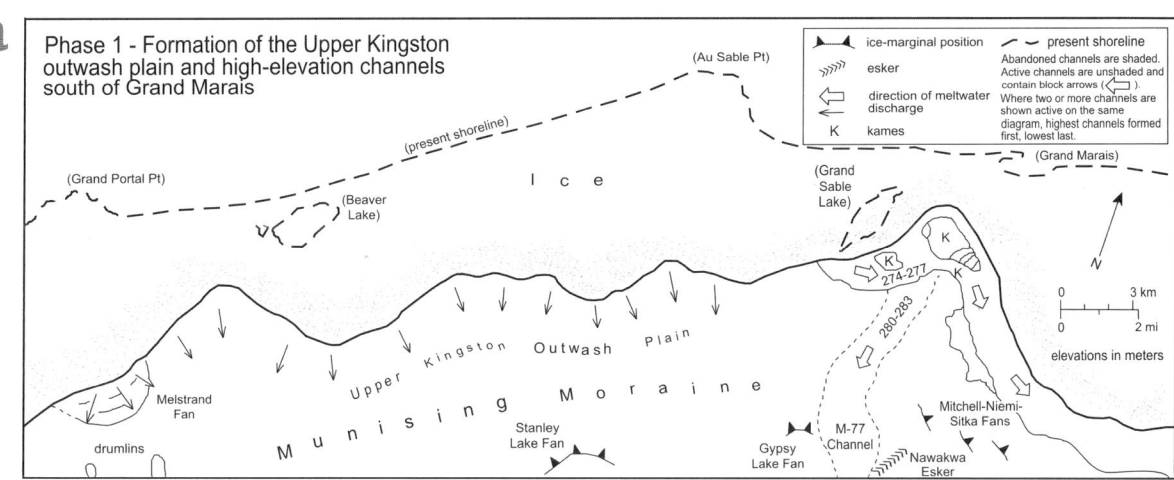

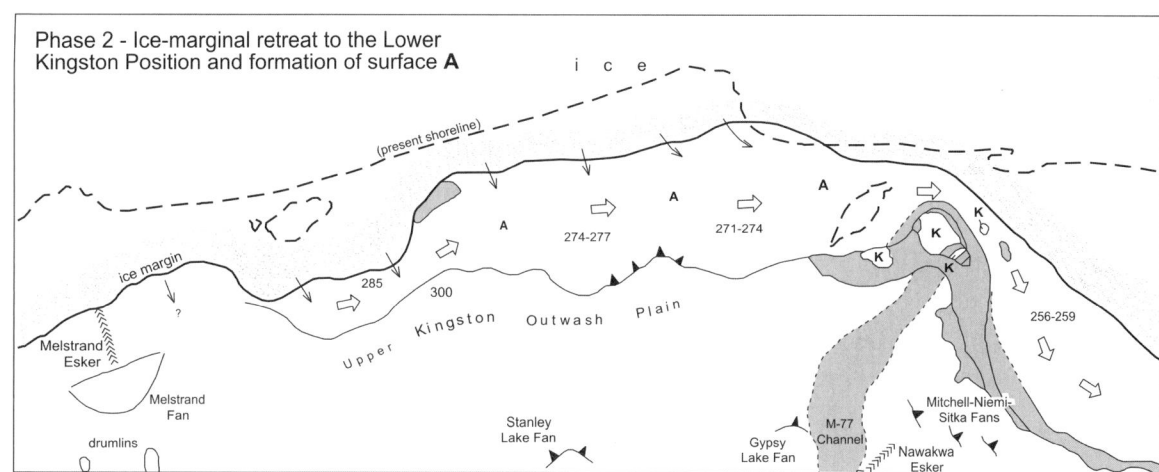

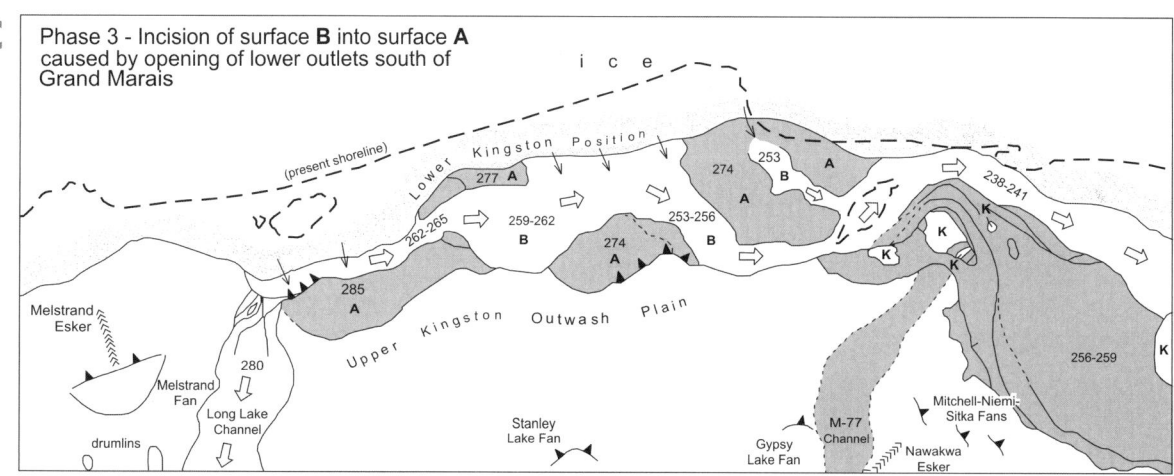

FIGURE 4.8 a, b, c. Deglaciation history of Pictured Rocks National Lakeshore (from Blewett, 1994).

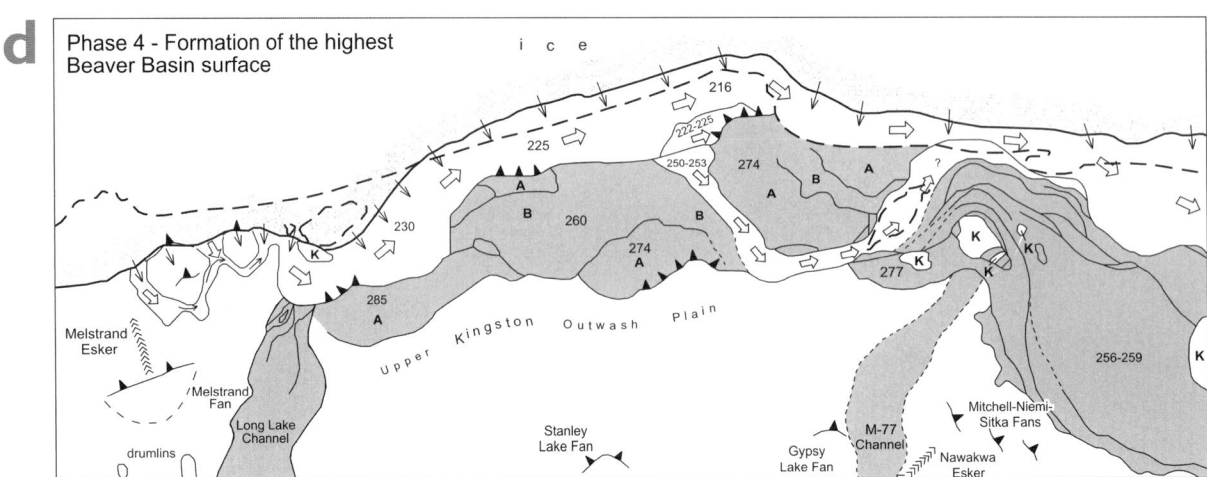

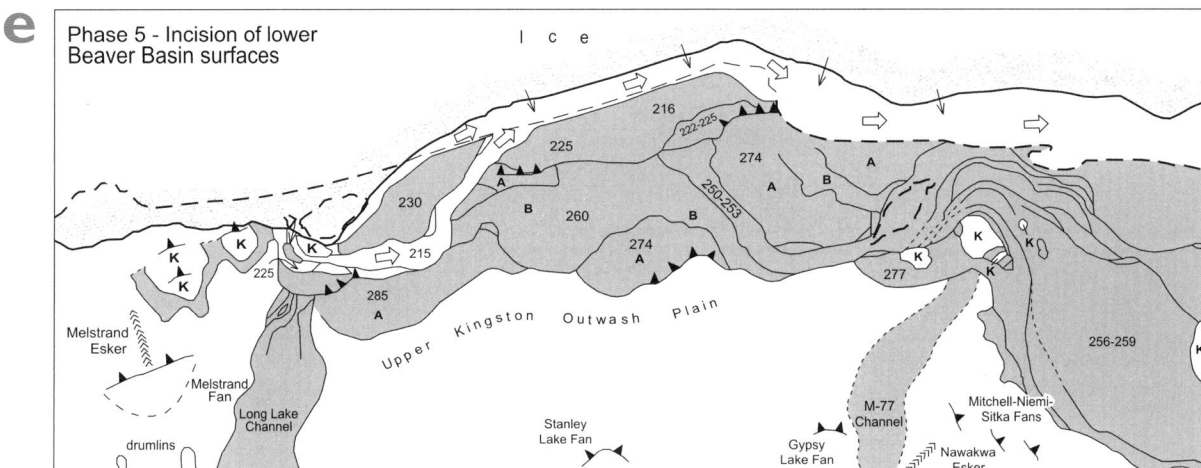

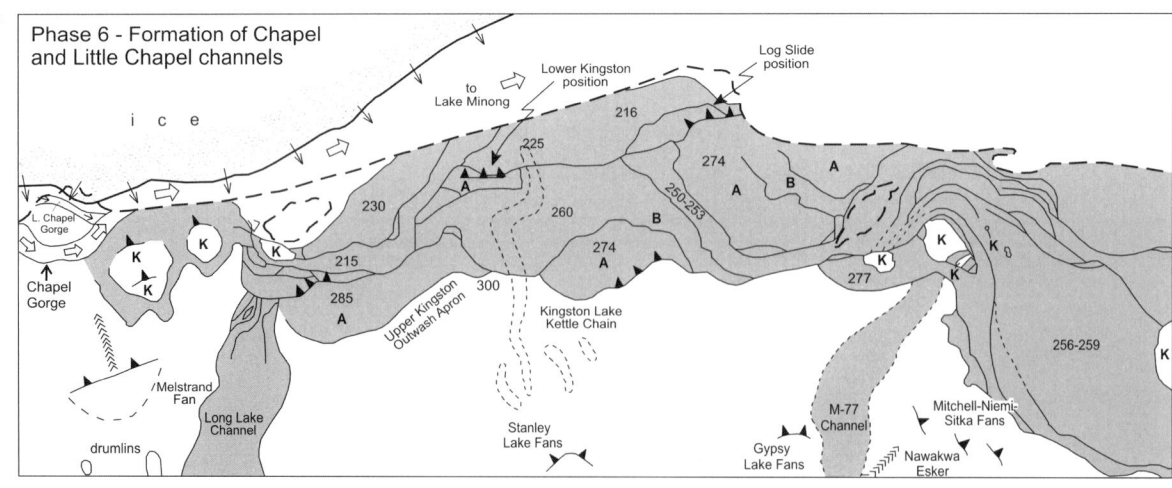

FIGURE 4.8 d, e, f. Deglaciation history of Pictured Rocks National Lakeshore (from Blewett, 1994).

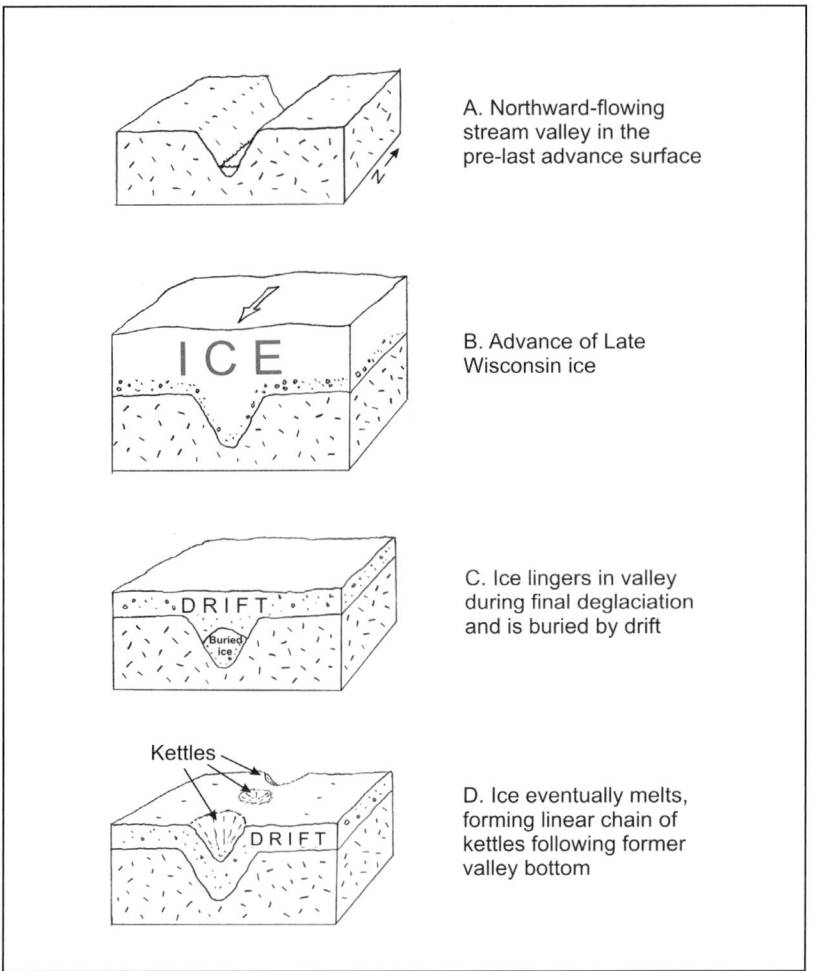

A. Northward-flowing stream valley in the pre-last advance surface

B. Advance of Late Wisconsin ice

C. Ice lingers in valley during final deglaciation and is buried by drift

D. Ice eventually melts, forming linear chain of kettles following former valley bottom

FIGURE 4.9. Formation of the Kingston Lake kettle chain (after Blewett, 1984).

can be observed for several miles along H-58 between the Miners Castle Road and the village of Van Meer (see road log).

Kame Terraces

The northern flank of the Munising moraine exhibits a complicated assemblage of kame terraces and incised channels formed by eastward flowing meltwater streams that were confined between the glacier's margin and ice-free highlands to the south (Figs. 4.7, 4.8b–f; see also Fig. 3.11). West of Beaver Lake, many channels were carved into bedrock, as at Chapel Lake gorge, but eastward the landforms are mainly constructional, including collapsed heads of outwash containing thick drift. The Grand Sable Banks near Grand Marais are developed upon thick accumulations of sand and gravel associated with these kame terraces. The terraces are best viewed at the Sable Falls parking lot at the eastern end of the park (see Figs. 3.11, 3.12; see road log).

Spillways between Lakes Superior and Michigan

In at least two instances, meltwater lakes dammed between the melting glacier and the Munising moraine to the south, spilled southward in catastrophic fashion. The Long Lake channel south of Beaver Lake (see Fig. 1.2a; Fig. 4.8c) is the most well developed, exhibiting an extensive 918-foot surface, with at least two smaller incised channels at 909 feet and 899 feet. Gravel pits along the channel bottom contain rounded boulders up to 1 foot in diameter in sand and coarse gravel, indicating a high velocity. The 899-foot channel also formed a distinctive meander scar, much like a modern river. By calculating the wavelength of the meander scar, geologists can determine the approximate paleodischarge of the meltwater stream that formed it (Dury, 1965). These calculations yield mean annual discharges of 1,100 to 1,200 m3/sec, or 45 times the annual discharge of the modern Tahquamenon River. Although large by modern standards, these figures are comparable to glacial meltwater flows from outwash features described by others (Clague, 1975; Maizels, 1983; Blewett, 1990). A second, less conspicuous, channel at approximately 928 feet parallels state highway M-77 between Seney and Grand Marais (Figs. 4.7, 4.8a; see Fig. 1.2b).

PREVIOUS WORK

Our current understanding of the glacial history of Pictured Rocks is based upon nearly a century of detailed study. The review that follows is provided for geologists and related academics who might require a more detailed understanding of the Quaternary geology. An attempt has been made to simplify and define concepts for the casual reader, however, wherever possible.

Frank Leverett (1911, 1917, 1929) mapped and interpreted the surficial geology of Michigan's Northern Peninsula, identifying the yet unnamed Munising feature as a moraine "deposited in water or later covered by waters of glacier lakes," and the flanking terraces west of Grand Marais as "sandy beds of glacial lakes" (Leverett, 1929, Plate 1). This mapping reflected the mistaken belief that after formation, most of the moraine was submerged beneath Lake Algonquin, an ancestral high-level lake that occupied the basins of Lakes Huron, Michigan, and Superior following final deglaciation (see chapter 3). J. P. Farrell and J. D. Hughes state (1985, p. 75):

> Working without the benefit of aerial photographs and topographic maps, Leverett misinterpreted many glacial features such as ice contact slopes and outwash channel banks as ancient shorelines. This led him to postulate that the high level Algonquin stage occupied not only the basins of lakes Huron and Michigan, but also the entire basin of Lake Superior.

S. G. Bergquist (1936b), following Leverett, described the moraine in detail. In a separate report, Bergquist (1936a) proposed that the Grand Sable Banks formed as a baymouth bar in Lake Algonquin, with Grand Sable Lake marking a remnant lagoon enclosed by the bar. H. M. Martin (1957) mapped the Northern Peninsula's surficial formations, closely following Leverett and Bergquist. J. L. Hough (1958), however, proposed that Lake Algonquin was restricted from most of the Lake Superior basin, and that outwash aprons of the Munising moraine were graded (or adjusted) to a lake in the Michigan and Huron basin that was slightly lower than Lake Algonquin and formed after the Main Lake Algonquin phase had ended. The term graded in this sense implies that the features involved are roughly coincident in time.

Geographer John Hughes of Northern Michigan University began research in the area in 1958 and was possibly the first to recognize the glaciofluvial origin of terraces and incised channels north of the moraine. He later confirmed their eastward slope and proposed a generalized glacial history that was incorporated into the Pictured Rocks National Lakeshore Master Plan (Hughes, 1968). He also addressed the significance of the Au Train–Whitefish Channel, an important relict drainageway between the Superior and Michigan lake basins, located west of the park (Hughes, 1989). Later, C. W. Drexler (1975) presented a deglaciation history similar to Hughes in most respects.

By 1975 geologists had developed a complex Great Lakes deglaciation chronology based in large part on radiocarbon dating. They concluded that beginning about 13,600 years ago the ice margin rapidly retreated northward from a terminal position near Two Rivers, Wisconsin. This recession was punctuated by two major still stands of the ice as represented first by the Newberry moraine, an east–west trending highland paralleling the Northern Peninsula's southern shore, and then the Munising feature (Fig. 4.6). Outwash aprons of the Munising feature appeared to terminate southward in a broad delta at the level of Lake Algonquin, an important and extensive glacial lake that occupied the Lakes Michigan and Huron basins (and part of the Superior basin) during initial deglaciation. P. F. Karrow and others (1975) dated Main Lake Algonquin to 13,000 years ago. If outwash aprons of the Munising moraine were graded to post-Algonquin levels as suggested by Hough (1958), then the age of the Munising feature was slightly younger than 13,000 years. Futyma (1981) contended that the Munising outwash was actually graded to the higher Main Algonquin level, placing the feature at 13,000 years. Thus, landform relationships between the Munising feature and ancient shorelines along its margin implied that the Munising moraine formed sometime around 13,000 years ago, or just slightly after.

Meanwhile, excavations at Gribben Lake near Marquette, Michigan, uncovered spruce stumps buried by outwash of the Marquette moraine (Hughes, 1978), a western extension of the Munising moraine (Fig. 4.6; Hughes, 1971). Five dates from wood samples provided an age of approximately 11,500 years. Based on these age estimates, Hughes named the ice-free interval of forest growth the Gribben Interstadial and the subsequent period of ice readvance the Marquette Stadial. Geologists use the terms stadial and interstadial to refer to significant readvances and withdrawals of the ice margin, respectively. Because the Marquette moraine was now firmly dated at approximately 11,500 years old, the correlative Munising moraine (and associated landforms) was assumed to be the same age, or about 1,500 years younger than previously thought. Hughes was unable to determine how far the ice margin had retreated into the Superior basin during the Gribben Interstadial before readvancing, but researchers studying Lake Agassiz father west would soon answer this question.

Detailed mapping of Lake Agassiz deposits and landforms by various workers (Clayton, 1983; Teller and Thorleifson, 1983) indicated that Lake Agassiz drained into the Lake Superior basin during two distinct episodes: the Moorhead (12,900 to 11,500 years ago) and Nipigon phases (10,600 to 9,300 years ago) (Teller and Thorleifson, 1983). The latter phase was the episode described in the last chapter, in which the drift barrier at Sault Ste. Marie may have been swept away. The location of Lake Agassiz outlets required that the Lake Nipigon area, and by analogy most of the Superior basin, be ice free during episodes of Lake Agassiz overflow. Because the earlier Moorhead phase coincided with Hughes's Gribben Interstadial, researchers concluded that the ice margin withdrew to the northern edge of the Superior basin between 12,900 and 11,500 years ago, allowing overflow from Lake Agassiz (Moorhead phase) to enter the then mainly ice-free Superior basin. Ice then readvanced southward during the Marquette Stadial, blocking Lake Agassiz outlets and building the Munising-Marquette moraines about 11,500 years ago. Eventual ice-marginal retreat reopened the Lake Superior basin about 10,600 years ago, allowing overflow from Lake Agassiz (Nipigon phase) to reenter the basin, possibly cutting the Sault Ste. Marie drift barrier.

Simply stated, the discoveries at Gribben Lake nicely complemented those associated with Lake Agassiz and indicated that the ice margin had paused along the northern edge of the Upper Peninsula approximately 13,000 years ago during its initial retreat, and then readvanced to the same general position 1,500 years later, before retreating for the last time. During each of these withdrawals, meltwater flowed into the Lake Superior basin, perhaps catastrophically, from Lake Agassiz.

These relationships mean that the Munising moraine's revised age is incompatible with the earlier proposals of Hough (1958) and Futyma (1981) that the feature's drainage was graded to Algonquin water levels. Specifically, a moraine

11,500 years old could not be graded to a lake that had drained 1,000 to 1,500 years earlier.

Drexler (1981) documented the sequence of lakes and drainageways (discussed in chapter 3) associated with ice withdrawal following the Marquette Stadial. Drexler and others (1983) further complicated the issue by proposing that the Munising moraine did not represent a significant stillstand of the ice margin, but was instead simply higher bedrock mantled with thin drift. In its place, a new feature, the Grand Marais moraine, was delineated and interpreted as having formed during the Marquette Stadial readvance. Drexler's Grand Marais moraine was thus somewhat younger than his discredited Munising moraine, which was interpreted to be a result of higher bedrock, not ice-marginal deposition. Drexler reiterated these interpretations in a later paper (Farrand and Drexler, 1985).

Blewett and Rieck (1987) questioned Drexler's conclusions and, based on topographic and subsurface data, showed that mean drift thickness in sections of the Munising moraine averaged 50 feet and likely exceeded 65 feet. In addition, they proposed that large parts of the moraine were complex ice-wastage features containing stratified drift that delimited three successive ice-marginal positions.

Later, Blewett (1994) mapped the outwash terraces along the Munising moraine's northern flank and proposed a revised deglaciation history. He also addressed the Munising moraine's apparent gradation to Lake Algonquin, proposing that the ice margin paused during initial retreat about 13,000 years ago, building large outwash aprons graded to Lake Algonquin water levels, and then readvanced to the same general location during the Marquette advance. Outwash aprons from the latter advance were apparently graded to much lower lakes in the Lake Michigan basin. These ideas were reiterated in a later paper (Blewett, 2002a).

T. V. Lowell and others (1999) confirmed and revised age estimates from the Lake Gribben forest bed. Their analysis yielded an age slightly older than the original estimates reported by John Hughes (1978). They also traced the margin of the Marquette advance for 600 miles along Lake Superior's southern shore and attributed the advance to climatic cooling during a widely recognized interval of climatic deterioration known as the Younger Dryas.

T. G. Fisher and R. L. Whitman (1999) studied sediment cores from Beaver Lake in Pictured Rocks National Lakeshore. Wood from the base of one core yielded a date of $9,480 \pm 60$ years B.P., corresponding to a calendar age of approximately 10,700 years, which they considered a minimum age for deglaciation of the area. Assuming that ice withdrawal from the Munising moraine began approximately 11,400 years ago (Lowell and others, 1999), their age estimate indicates that the retreating ice margin had reached an unknown position north of Beaver Lake after approximately 700 years.

As of this writing, the weight of evidence suggests that terraces and incised channels north of the moraine within Pictured Rocks National Lakeshore are likely associated with ice-marginal retreat following the Marquette Stadial, making most glacial landforms within the park approximately 10,700 to 11,500 years old. The Munising moraine may be a compound feature recording at least two phases of glacial deposition.

More recently, Bob Regis of Northern Michigan University and his colleagues (Regis and others, 2003; Regis and others, 2004; Derouin and Regis, 2005) studied a large branching network of north–south oriented valleys on the floor of Lake Superior, just north of the Pictured Rocks area. They interpreted these features as tunnel channels cut by pressurized meltwater at the base of the glacier and proposed that some of these channels served as meltwater sources during formation of the Upper Kingston Outwash Plain.

DEGLACIATION HISTORY

As stated previously, the Wisconsin glacier had by 13,600 years ago reached a position near Two Rivers, Wisconsin. From here, the ice margin continued its withdrawal northward, forming the Newberry moraine along the southern shore of the Upper Peninsula and then, by about 13,000 years ago, depositing the broad heads of outwash associated with the Munising moraine along the park's southern boundary. These features were probably graded to Lake Algonquin or a slightly later post-Algonquin lake phase in the Lake Michigan basin. Continued retreat of the margin to a position north of the present shoreline ensued during the Gribben Interstadial about 12,400 years ago, followed by a readvance during the Marquette Stadial 11,500 years ago. The recently abandoned Munising moraine, perched atop the underlying Cambro-Ordovician bedrock escarpment, likely formed a significant barrier to southward advance of the ice, and a new series of heads of outwash and related features were deposited atop the Munising moraine. These features were likely graded to much lower lake levels in the Lakes Michigan and Huron basins.

At this point in the story, the formation of the various glacial landforms in Pictured Rocks National Lakeshore and vicinity can finally be addressed. The six-phase deglaciation history described here is based on detailed work of the present author (Blewett, 1994), following Hughes (1968). Elevations are presented in meters in figure 4.8 following data from the original report. Corresponding elevations are provided in feet. Detailed township and range designations for many of the key landforms are provided in the 1994 report. Because the terminal position of Marquette ice cannot be determined with certainty, the glacial history begins with the first late Wisconsin ice-marginal positions to have influenced topography within Pictured Rocks National Lakeshore.

Phase 1

Sometime after 11,500 calendar years, the ice margin likely stood along the Melstrand fan and Upper Kingston outwash plain (Fig. 4.8a). Supporting evidence includes boulder concentrations along apron crests south of Beaver Lake. Meltwater flowed south from this position as indicated by southward-fining surficial deposits. The great size and extent of these features suggests formation during an equilibrium period of significant duration, perhaps lasting several decades. As ice withdrew from this position, meltwater dammed between the glacier border and the adjacent highlands spilled southward, forming a wide channel at 283 meters (928 feet) that today parallels highway M-77 (M-77 channel, Fig. 4.8a). Continued retreat opened a lower channel at 274 to 277 meters (899 to 909 feet) south of Grand Marais (Fig. 4.8a). Several higher ancillary surfaces are located nearby.

Phase 2

Ice-marginal retreat to a new position along the Lower Kingston plain (Hughes, 1968) opened lower eastern drainage outlets. Meltwater, confined between the terminus and the now abandoned Upper Kingston ice-contact slope, formed a broad, eastward-sloping kame terrace heading at approximately 285 meters (935 feet) (surface A; Fig. 4.8b). Boulders on the crest indicate the presence of ice-marginal positions northwest of Kingston Lake and west of the Log Slide.

Phase 3

While the glacier's border remained at the Lower Kingston position, marginal retreat from the highland south of Grand Marais opened a still lower outlet, permitting meltwater streams to incise a new channel (surface B) into surface A (Fig. 4.8c). Further recession allowed incision of a lower, 250- to 253-meter (820- to 830-foot) channel that today contains the Hurricane River (Fig. 4.8d). These latter surfaces likely record the transition to nonequilibrium conditions in the glacier just prior to abandonment of the Lower Kingston position.

Meanwhile, meltwater trapped between the ice margin and the Lower Kingston ice-contact slope south of Beaver Lake spilled southward, forming the Long Lake channel (Fig. 4.8c). This feature exhibits an extensive 280-meter (918-foot) surface, with at least two smaller incised channels at 277 meters (909 feet) and 274 meters (899 feet).

Phases 4 and 5

Eventually, ice withdrew to a position several kilometers north of the Lower Kingston margin, initiating eastward meltwater drainage across a broad terrace heading at 230 meters (755 feet) (Fig. 4.8d). Further retreat initiated two episodes of incision at 225 meters (738 feet) and 215 meters (705 feet) into the 230-meter channel (Fig. 4.8e). These surfaces are best observed south of Beaver Lake.

Phase 6

The glacial border finally retreated to a position just south of Grand Portal Point (Fig. 4.8f), forming the last barrier separating lakes in the western and eastern sections of the Lake Superior basin (Drexler, 1981; see Fig. 3.8a). Further ice withdrawal allowed western lakes to spill eastward into an early version of Lake Minong, cutting the Chapel and Little Chapel gorges. Continued ice-marginal retreat allowed Lake Minong to expand north and westward into the remaining parts of the basin by 10,700 years ago, to eventually form the first postglacial lake in the Superior basin (Farrand and Drexler, 1985).

DISCUSSIONS AND CONCLUSIONS

Surficial sediments and landform assemblages differ markedly between western and eastern sections of the Pictured Rocks region. These may be due to changes in the behavior of the glacier during deglaciation. Initially, the bottom of the glacier may have been frozen to its bed, a type of glacier geologists call a cold glacier. As melting continued, conditions may have changed to those of a warm glacier, in which the glacier is not frozen to its substrate. The Melstrand esker and small, incised, subglacial channels west of Miners River are indicative of well-developed subglacial drainage systems commonly associated with warm-based glaciers in which surface meltwater can penetrate to the glacier bed (Mooers, 1990). In contrast, large coalescing outwash aprons (Upper Kingston apron, Section 35 Hill) and hummocky ice-wastage topography form from large accumulations of superglacial drift (brought to the surface by shearing within the ice) that may be associated with colder glaciers with frozen margins (Mickelson and others, 1983; Mooers, 1990). Collectively, these various landform assemblages may indicate a change from cold- to warm-based conditions during final deglaciation, typical of retreating glaciers (Clayton and Moran, 1974). Thus, differences among landforms and surficial sediment characteristics in the region likely represent contrasting spatial and temporal variations in thermal conditions of the ice sheet.

Most glacial terrain within Pictured Rocks National Lakeshore and vicinity formed within a 700-year period commencing approximately 11,400 years ago as a retreating ice margin confined meltwater streams against the Munising moraine to the south. Spatial and temporal differences in glacier dynamics likely produced the significant variations in surficial sediment characteristics and landform assemblages present. Large coalescing outwash aprons, kame terraces, and incised channels may record long intervals of quasi-stable ice-marginal conditions favoring thick outwash deposition, punctuated by comparatively brief transitional periods of rapid meltwater incision due to ice-marginal retreat from highlands south of Grand Marais. The Munising moraine represents a significant ice-marginal accumulation containing abundant ice-contact and stratified drift formed along a stagnant ice margin.

5

THE HOLOCENE EPOCH

T he Holocene history of Pictured Rocks begins with the final wastage of
Marquette ice into the Lake Superior basin approximately 11,000 years ago.
As the land emerged from beneath the glaciers, weathering, mass wasting,
and erosion began to reshape the landscape, although with limited impact.
Indeed, the profound changes wrought by glacial conditions were not easily erased,
and a long transition period of crustal, topographic, and lake level adjustment ensued
(see chapter 3), which continues in diminished form today. By 5,000 years ago,
about the time of the first pyramids in Egypt, nearly all of the essential components
of the modern landscape were in place, although in slightly different forms than
today, including the Pictured Rocks cliffs and Grand Sable Dunes. From a long-term
geologic perspective, however, the manifest processes operating during the Holocene
have had little effect on the region's overall surface configuration, especially away
from the coast, and present topography can be viewed as a relict form, little changed
from the latest Pleistocene.

A chronological approach to Holocene events is difficult because many of the
most significant events were occurring simultaneously. This chapter addresses
Holocene events from four perspectives: formation of the modern network of
streams and lakes, initiation and evolution of the Grand Sable Dunes, maintenance of
the Pictured Rocks cliffs and related coastal landforms, and human influences on the
modern landscape. The associated glacial and postglacial lake sequence that extends
into the Holocene has already been discussed in chapter 3, but serves as the basis for
some of the discussions that follow.

STREAM AND LAKE NETWORKS

Streams

As the Marquette ice margin retreated from the region, Lake Minong, the lake
occupying the southeastern corner of the Superior basin, began to expand north

and westward, eventually forming the first postglacial lake in the Superior basin (see Fig. 3.8c). With their meltwater source extinguished, the eastward flowing braided outwash streams that had formerly drained the glacial margin dried up, or nearly so, and were replaced by smaller single-channel streams controlled by groundwater and precipitation. Initially, these rivers continued to occupy the abandoned kame terraces, and drainage was to the east. Near the shore of Lake Minong, however, surface runoff quickly formed short stream segments perpendicular to the shoreline. These streams had steep profiles, allowing them to expand their tributaries upstream with time, eventually capturing the drainage of streams still flowing along the terraces in a process called stream piracy. Thus, within just a few centuries, surface streams likely changed from extensive eastward flowing braided channels dominated by glacial meltwater to smaller single-channel streams controlled by groundwater and precipitation. Stream piracy, working over the ensuing millennia, favored the expansion of northward flowing over eastward flowing streams. This process is becoming less important today as the southern shore of Lake Superior is slowly drowned due to continued isostatic adjustments (chapter 3).

Lakes

Within the first few thousand years of postglacial conditions, buried ice blocks began to melt from beneath the drift, forming the Kingston Lake kettle chain and the thousands of related kettles found throughout the region. Where the water table was high enough to intersect the bottoms of these kettles, kettle lakes such as Kingston and Turtle Lakes (see road log) were formed. Even today many lakes on the Kingston Plains, including Kingston Lake, markedly expand and shrink seasonally with fluctuations in the groundwater table. Coherent river networks connecting and draining these lakes were slow to materialize, however, due to the permeability of the underlying sands, high water table, and low relief.

Other lakes formed in different ways. Chapel and Little Chapel Lakes occupy low spots carved into bedrock by former meltwater channels. Grand Sable Lake is a special case all its own, and is addressed in the next section. Beaver, Little Beaver, and Trappers Lake represent remnants of former lagoons in Lake Nipissing, which were eventually separated from the main lake by bay mouth barriers and falling Nipissing lake levels. Data from cores drilled into the bottom of Beaver Lake by T. G. Fisher and R. L. Whitman (1999) record a complicated series of events related to falling (Houghton low phase) and rising (Nipissing transgression) lake stages in the Superior basin during the early and middle Holocene. Their study provides valuable data for understanding the formation and evolution of these lakes.

INITIATION AND EVOLUTION OF GRAND SABLE DUNES

The Grand Sable Dunes cover more than four square miles and are one of the largest perched dune complexes in the Great Lakes region (Fig. 5.1). The dunes are perched dramatically atop the massive Grand Sable Banks, a 100- to 300-foot-high bluff that dominates the Lake Superior coastline from Au Sable Point eastward to Grand Marais (see Fig. 1.4). The dunes cover at least two extensive kame terraces at approximately 880 and 720 feet in elevation, whose northern edges have been truncated by modern Lake Superior (Blewett, 1994). Thus the lower and middle portions of the bluffs are composed of outwash and related deposits containing abundant gravel, sand, and silt, with dune sand dominating the upper section. This fact is often discovered the hard way by visitors descending the bluff in leaps and bounds from the Log Slide overlook. The dune field is most active on its eastern side, where the bluff is destabilized by wave action from Lake Superior. Farther west, the Grand Sable Banks are often protected from the waves of the open lake by Au Sable Point, and the dunes diminish. The Grand Sable Dunes are bordered on the southeast by Grand Sable

FIGURE 5.1. Oblique aerial view of Grand Sable Dunes. View looking eastward. (Courtesy of National Park Service)

Lake and its outflow to Lake Superior, Sable Creek (see Fig. 3.11). Many researchers have remarked upon the curious juxtaposition of these landforms, and the Holocene history of this area has sparked significant debate.

Previous Work

As mentioned in chapter 4, Stanard Bergquist (1936a) was the first scientist to carry out detailed fieldwork in the dunes, proposing that the Grand Sable Banks formed as a bay-mouth bar in Lake Algonquin, with Grand Sable Lake marking a remnant lagoon enclosed by the bar. Geographer John Hughes worked in the area during the 1960s and '70s, joined later by his colleague Pat Farrell. In a report commissioned by the National Park Service, Farrell and Hughes (1985) described the Grand Sable area in great detail, establishing the seminal ideas that would inspire later workers. Their report also marks the first mention of an extensive and well-developed buried soil, or paleosol, found in Grand Sable Dunes that would later be called the Sable Creek soil (Fig. 5.2; see Fig. A.18 in the road log). Paleosols are soils that were once a stable ground surface but have been subsequently buried beneath accumulated sediments. Those developed in outwash often begin as a gravel "lag" on the surface, as wind winnows away sand, leaving the heavier gravel behind. With time, the surface

FIGURE 5.2. The Sable Creek soil, a prominent paleosol in the Grand Sable Dunes area. (Courtesy of Walt Loope, U.S.G.S.)

becomes armored with gravel and protected from further wind erosion, allowing vegetation to take hold. With continued exposure to the atmosphere, a thick soil profile (a kind of weathering rind on the surface) may develop upon the landscape. Normally, the longer the soil is exposed to the atmosphere, the thicker and better developed it becomes. If fire, climatic change, rising lake levels, or other disturbances cause the protective vegetative cover to be removed, and if abundant sand is available, prevailing winds may remobilize the sand fraction as dunes and the soils downwind are buried and preserved. Sometimes these dunes bury mature forests that, much later, may be reexposed by deflating winds. These ancient "ghost" forests, most of which date to the late Holocene, were visible from the Dunes trail near Sable Falls (see Fig. A.19 in the road log) until very recently, and may reappear in the future.

Farrell and Hughes (1985) recognized paleosols and lag deposits as important factors in dune dynamics in Grand Sable Dunes (p. 24–25):

As the North American ice sheet receded from the Grand Sable . . . the outwash sediments dried and began to move with the wind. Usually, on areas of outwash, which are inherently poorly sorted, the length of the period of wind erosion was limited by rather rapid accumulation of a lag (desert armor) consisting of clasts [rock pieces] in excess of one centimeter in diameter. This formation of a lag on the various surfaces at Grand Sable must have permitted only a very short period of initial instability when wind could rework sand into small dunes and possibly sand sheets. However, in contravention to the paucity of sediment of an eolian [wind] origin upon the Grand Sable that should be expected there is instead a great mass of wind blown sand over the surface ranging frequently to more than 70 feet thick. . . . Clearly, persistent undercutting of the Grand Sable Bank by wave erosion precludes the formation of a lag upon its north face. Thus, sand is constantly, freshly exposed to wind erosion and although most of it eventually slumps or slides into Lake Superior a portion is blown upslope, over the bluff crest, and onto the lag surface. The history of the Grand Sable must include periods when low lake levels with little or no wave erosion allowed the accumulation of a protective cover of stone upon the bluff. At other times, when higher lake stages occurred, wave erosion and bluff retreat must have lead to extensive wind transport of sand from the bluff onto the top of the Grand Sable.

Farrell and Hughes attributed the main phase of dune building to the rise and stabilization of Lake Nipissing (chapter 3) approximately 6,000 years ago. As Lake

Nipissing fell, a renewed period of stability and lag formation may have ensued and dune formation slowed. Establishment of the modern lake level approximately 2,300 years ago, however, led to renewed erosion along the base of the Grand Sable bluff, and the dunes began their present active phase. If correct, this conclusion means that the dunes date only to the middle part of the Holocene. It is perhaps a testament to the quality of their research that the basic conclusions of Farrell and Hughes have remained intact, despite nearly twenty-five years of subsequent research in the Grand Sable region.

During the 1970s and '80s, Bill Marsh of the University of Michigan–Flint, aided by his brother Bruce Marsh of Johns Hopkins University, studied the dunes using a quantitative approach focused on wind processes (Marsh, 1990; Marsh and Marsh, 1987). They concluded that the bluff's upper face is indeed the source for most of the sand making up the Grand Sable Dunes and calculated an annual sand nourishment rate of 6,300 m3 over a 10-year period. Using this figure, they concluded that a middle Holocene age for the dunes as proposed by Farrell and Hughes could only account for about half of the volume of sand ($68 \times 106 \text{ m}^3$) present in the modern dune field. They suggested that a rising Lake Minong might have initiated dune formation as early as 10,100 years ago, making the dunes early Holocene, or about 5,000 years older than proposed by Hughes and Farrell.

In the early 1990s, geographer John Anderton began work on the buried soils of the Grand Sable area, aided by Walt Loope of the National Biological Service. Loope had come to Pictured Rocks National Lakeshore in 1982, first as a resource management specialist for the park and then as an ecologist. With the "reinventing government" initiative of the 1990s, Loope was reassigned to the National Biological Survey, which later was folded into the U.S. Geological Survey. During all these transitions, Loope remained headquartered at Pictured Rocks and had a profound influence on subsequent research in Grand Sable Dunes.

Anderton and Loope (1995) named the extensive well-developed paleosol recognized by Farrell and Hughes (1985) the Sable Creek soil. They mapped and described it in detail, noting that the soil was developed upon kame terraces made of outwash (Fig. 5.2). They also recognized numerous poorly developed paleosols above the Sable Creek soil, some containing stumps and other datable material that they attributed to at least four (and possibly eleven) separate episodes of dune building. Radiocarbon dates and other data indicated that the Sable Creek soil was buried between about 6,000 and 5,350 years ago, dates that nicely coincided with the rise of Lake Nipissing, which reached its maximum between 5,400 and 4,775 calendar years. Based on these data they concluded that:

During the middle and late Holocene, the Grand Sable Dunes appear to have experienced episodic stability and instability, mediated by the condition of the lakeward face of the banks as it responded to fluctuations in the level of Lake Superior. The dunes appear to have originated during the Nipissing Great Lakes after [6,800 years ago] and experienced episodes of increased eolian activity that buried soil profiles and forests at about [6,000; 5,300; 3,775; and 525 years ago]. . . . Episodes of eolian activity were separated by periods of relative stability that allowed forest growth and soil profile development. (p. 196–197)

Thus, in a very simple sense, when lakes were rising, the dunes were active, and when lakes were lowering, they stabilized. The nature and extent of the Sable Creek soil also provided strong support for the contention that the Grand Sand Dunes dated to the mid-Holocene and were linked to the rise and stabilization of Lake Nipissing. According to this reasoning, the Sable Creek soil must have begun forming on outwash soon after the glacier melted and likely required several thousand years to form. The rising lake levels of the mid-Holocene caused the adjacent lake bluff to destabilize, burying the soil with sand. Thus, Anderton and Loope's work supported Hughes and Farrell's contention that the dunes dated to the mid-Holocene. This new and deeper understanding of dune dynamics would provide the catalyst for the next phase of the Grand Sable story, one that would take some unexpected turns.

The Curious History of Sable Creek, Grand Sable Lake, and Grand Sable Dunes

Grand Sable Lake occupies the bottom of an incised southwest–northeast trending former meltwater channel at approximately 743 feet in elevation (Hughes, 1968; Blewett, 1994). It is fed by three small creeks along its southern and eastern margin and drains along the southeastern edge of Grand Sable Dunes via Sable Creek to Lake Superior (see Fig. 3.11). Previous researchers have interpreted the lake as a former lagoon in Lake Algonquin (Bergquist, 1936a) and as a large kettle (Farrell and Hughes, 1985).

During the winter of 1999, Tim Fisher, then at Indiana University–Northwest, and Walt Loope took several sediment cores through the ice of Grand Sable Lake. Because the lake was located downwind from the dunes, they reasoned that the lake would have served as a natural sink for blowing sand when the dunes were active, forming thick accumulations of sand on the lake bottom (called rhythmites), with thinner layers deposited during inactive phases. Instead, their cores revealed the remnants of drowned forest floors, including several in-place submerged stumps. Radiocarbon dates indicated that drowning had occurred at least three times at 3,000,

1,530, and 300 years ago (Loope and others, 2004). Thus it appeared that Grand Sable Lake, at least in its present form, was not a permanent feature but had expanded and contracted multiple times during the Holocene.

What could cause this cyclical drowning and draining of the lake? As the natural outlet for Grand Sable Lake, Sable Creek was immediately suspected. Blocking the creek would cause Grand Sable Lake to back up behind any obstruction. If the obstruction were somehow removed, the lake would shrink as it drained to lower levels. Suddenly, the curious geography of Sable Creek and the Grand Sable Dunes began to make sense. Even a small southeastward advance of an active Grand Sable Dunes could block Sable Creek and dam the lake. Better yet, the timing of the drowning events seemed to generally coincide with the periods of active dune formation reported by Anderton and Loope (1995) based on data from paleosols within the dunes. Before any such proposal could be accepted, however, more evidence was needed.

Loope and his colleagues realized that Sable Creek was the key to the story, and they began a detailed map and field inventory of the regional drainage network associated with Sable Creek. It soon became apparent that the drainage history was much more intricate than first realized, characterized by a complicated sequence of stream piracies and drainage diversions likely dating back more than 6,000 years. Stranger yet, it appeared that a major paleovalley of Sable Creek lay buried beneath the Grand Sable Dunes.

Harry Jol of the University of Wisconsin–Eau Claire had confirmed this valley using a geophysical device called Ground Penetrating Radar (GPR). When used carefully, GPR can provide high-resolution images of sedimentary structures lying at depths up to 150 feet below the surface. Jol, a recognized expert on GPR, had been invited to the Upper Peninsula to help in the Grand Sable Lake research. He had only recently participated in a GPR project along the Columbia River delta in Oregon, searching for the remnants of a native village reported by Lewis and Clark. The Grand Sable region was an excellent field site for use of GPR, which works best in sandy sediments like dunes and outwash.

Based on several GPR transects across the dunes, the paleovalley of Sable Creek could be traced from the northern edge of Grand Sable Lake northwestward beneath the dunes to the Lake Superior bluff (Fig. 5.3), and it was clear that this former outlet of Grand Sable Lake had been buried deeply by Grand Sable Dunes. East of modern Sable Creek, a number of similar dry channels were exposed at the surface and could be traced along the stair-step-like series of kame terraces near the village of Grand Marais. Some of these former streams had been pirated by First Creek, a deeply incised northward flowing stream located just to the east of the park boundary (see

Fig. 3.11). A particularly fine example of one of these pirated paleovalleys can be observed via the hiking trail at Donahey Woods in Grand Marais (see Fig. A.23) and from trails at Sable Falls (see road log).

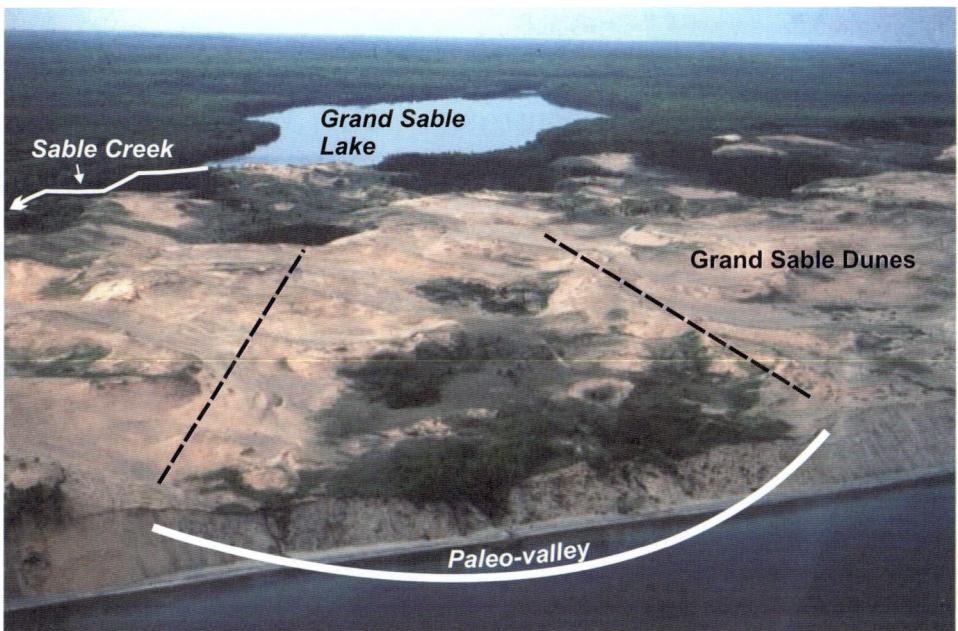

FIGURE 5.3. View southward towards Grand Sable Lake (in distance), showing the approximate location of the Sable Creek paleovalley (from Loope and others, 2004). (Courtesy of Walt Loope, U.S.G.S.)

Based on all this evidence, Loope and his colleagues constructed a complicated story of diversions, piracies, and dune advances associated with the damming and draining of Grand Sable Lake (Loope and others, 2004). They concluded that repeated encroachment of the dunes during rising lake levels diverted the channel of Sable Creek progressively eastward along the trend of the abandoned kame terraces before finally reaching Lake Superior (see Fig. 3.11). As lake levels dropped after about 5,300 years ago, northward flowing streams with short drainage to the coast were invigorated and captured the drainage of Sable Creek, resulting in a steady westward progression of Sable Creek back into the now stabilized dune field. Another rise in lake level about 4,500 years ago activated the dunes once again, forcing Sable Creek back to the east. The details of the story are beyond the scope of this book. Indeed, as Loope and others (2004, p. 320) note:

> The circumstances that drove landscape changes in the vicinity of Grand Sable Lake during the Holocene are, perhaps, unusually complex. Evidence presented here, however, suggests that lake level-mediated dune building,

a relatively common phenomenon throughout the Holocene, . . . may have driven repeated hydrologic adjustment within many small watersheds along erosional segments of the upper Great Lakes coastal zones.

THE PICTURED ROCKS CLIFFS

Meanwhile, in western sections of the park, the rising waters of the Lake Superior basin began to undercut the Pictured Rocks escarpment, newly emerged from beneath the ice. Its precise position and configuration in the early Holocene is unknown, but it must have been located some distance (1/2 to 1 mile?) north of the present cliffs and may already have exhibited significant relief, perhaps modified by a drift cover of varying thickness. Appreciating the suite of processes that have shaped the cliffs since their inception is perhaps best achieved by examining present processes, which have operated more or less unabated since the mid-Holocene.

The Pictured Rocks cliffs are only found where the Cambro-Ordovician escarpment faces the open waters of Lake Superior. Wave action, especially during storms, is largely responsible for undercutting the cliffs, leading to massive collapses of the overlying rock along preexisting fractures. These fractures are maintained through a variety of mechanical and chemical weathering processes, most notably

FIGURE 5.4. Prominent fracture (arrow) aligned parallel to the Pictured Rocks cliff face.

freeze-thaw (Fig. 5.4). Fractures often align in particular directions in rocks in response to stresses imparted during and after their formation. W. K. Hamblin (1958) reports that at least one major set of parallel fractures in the Munising Formation is oriented southwest–northeast. This situation may explain the tendency for the best-developed cliffs to form where the modern coast parallels this general orientation, as between Sand Point and Grand Portal Point. Wave erosion by the modern lake, as well as by higher lake stages, has produced a variety of erosional coastal landforms including headlands, arches (Fig. 5.5), stacks, and caves (Fig. 5.6). Evidence for mass wasting along the cliffs is clearly indicated by the large piles of sandstone rubble located in the water along the base of the cliffs (Fig. 5.7).

FIGURE 5.5. Shoreline arch along the Pictured Rocks cliffs just east of Mosquito Beach.

FIGURE 5.6. Shoreline cave formed by undercutting of rock layers by wave action. (Courtesy of National Park Service)

The persistence and rapidity of this process is recorded in the numerous rock falls over the historic period, most notably the collapse of Grand Portal Point in 1900 (Fig 5.8). The largest recent event occurred on July 29, 2010, and was captured with a digital camera by professional photographer Lou Waldock from a Pictured Rocks cruise boat (Fig. 5.9). Another major collapse occurred on April 13, 2006, when a portion of Miners Castle toppled into the lake (Fig. 5.10). Smaller events occur nearly every year. Small dune fields often develop atop the cliffs after a rock fall, resulting from the deflation of newly exposed drift by the prevailing winds.

FIGURE 5.7. Undercutting of cliff faces, combined with freeze-thaw and other weathering processes, eventually lead to catastrophic rock fall and accumulated debris at the base of the cliff.

FIGURE 5.8. The collapse of Grand Portal Point in 1900. Top picture shows the cliff in the late nineteenth century before the rock fall. The bottom picture shows the cliff soon after the event. Nearly all of the collapsed debris has been removed or redistributed by the lake over the past one hundred years. Such rock falls initiate small and isolated perched dunes along the bluff edge that are analogous (in terms of process) to Grand Sable Dunes. These areas also support small patches of dune vegetation embedded within a "sea of hardwoods." (Courtesy of Gregg Bruff, National Park Service)

FIGURE 5.9. The spectacular July 29, 2010, rock fall just west of Spray Falls was captured by professional photographer Lou Waldock from a Pictured Rocks tour boat. (Copyright 2010 by Lou Waldock, used with permission)

The "pictures" in the Pictured Rocks come from mineral stains that occur as groundwater oozes out of the bedrock face, often along tiny cracks (Fig. 5.11). The dripping water contains iron, limonite, copper, manganese, organic minerals derived from rotting vegetation, and other compounds that are deposited as vertical streaks of color as the water trickles down the rock face. Iron stains are commonly red, limonite brown, algae and copper green, manganese and organic materials black. The colors are best seen by boat in the early evening when the setting sun illuminates the rocks from the northwest. No pictographs or petroglyphs have been observed on the rocks.

Where the Pictured Rocks escarpment trends away from the open lake, the cliffs die out and are covered by glacial and postglacial debris. This is observed most clearly south of Sand Point near Munising, where Grand Island protects the eastern side of Munising Bay from strong wave action. Munising Falls and several other smaller waterfalls (see Figs. A.1, A.2 in road log), have cut through the overburden, exposing the escarpment in places. These high bluffs surround the city of Munising and continue westward toward Marquette, where they become buried beneath thick drift. On its eastern margin, the Pictured Rocks cliffs end where the escarpment turns southward away from the lake just east of Grand Portal Point. It can be followed, however, in the forested bluffs south of Beaver Basin, where glacial meltwater has

FIGURE 5.10. Before (*left*) and after (*right*) photographs of Miners Castle. On April 13, 2006, the easternmost (and largest) turret collapsed into Lake Superior. (Left photo courtesy of National Park Service)

removed some of the overburden, and in a few small outcrops along the highlands south of Grand Marais. Farther east, the rocks are famously exposed at Lower and Upper Tahquamenon Falls, formed upon the Chapel Rock and Miners Castle / Au Train units, respectively.

OF AGATES AND ERRATICS

Many a rockhound has been born along the gravelly beaches of Lake Superior. These rocks are a complicated mixture of Paleozoic and Precambrian gravels (including a few Paleozoic fossils) that were eroded, transported, and dumped by the glacier and

FIGURE 5.11. "Pictures" in the Pictured Rocks cliffs. The various colors come from mineral stains that occur as groundwater oozes out of the bedrock face, often along tiny cracks.

FIGURE 5.12 a, b, c. Typical beach gravels found in Pictured Rocks National Lakeshore (a). Rounded granite pebbles (b) are common and distinguished by large crystals colored light gray, white, pink, or red, with about 15–20 percent black minerals. Epidote (c) exhibits an unusual black-green to yellow-green color. (5.12b and 5.12c courtesy of National Park Service)

later reworked by the waves of Lake Superior. Many of these rocks are exotic to the central Upper Peninsula and come from bedrock sources in Canada (Fig. 5.12a). A detailed description of the multitudes of beach stones found within the region is beyond the scope of this book. Interested readers are referred to Kevin Gauthier and Bruce Mueller's excellent *Lake Superior Rock Picker's Guide* (2007) and Dan and Bob Lynch's *Michigan Rock and Minerals* (2010) for further information. Agate hunters should consult Axel Niemi's *Michigan's Glacial Gemstones of* [the] *Northeastern Upper Peninsula* (1973, now out of print) and his friend and apprentice Karen Brzys's lively and colorful publication *Understanding and Finding Agates* (2004) available at the Gitche Gumee Agate and History Museum in Grand Marais (see road log). Susan Robinson's *Is This an Agate?* (2001) is a richly illustrated and accurate guidebook with detailed directions to the best beaches for collecting. Scott Wolter's *The Lake Superior Agate* (2008) is also highly recommended.

An agate is a translucent variety of quartz called chalcedony (chal-SED-o-nee) characterized by colors arranged in alternating bands, in irregular clouds, or in mosslike forms (Bates and Jackson, 1987, p. 10; Fig. 5.13). It can be of any color and forms in small openings called vugs in volcanic rocks or in cavities of other rocks.

FIGURE 5.13. Lake Superior agates. (Courtesy of National Park Service)

The processes involved in agate formation are still a matter of scientific debate (Macpherson, 1989; Shaub, 1979; Wang and Merino, 1990; Pabian and Zarins, 1994), but nearly all involve the precipitation of quartz-rich fluids within openings in preexisting rocks. Proterozoic basalts associated with the Lake Superior Rift Zone are believed to be the source rock for many Lake Superior agates (Brzys, 2004). Although agates can be polished into handsome semiprecious gemstones, finding them amid the confusion of a Lake Superior beach can be extremely difficult. Good spots for agate hunting are along the beach just west of the Grand Marais Tourist Park (see road log) and along the Superior coastline from Grand Marais Bay eastward to Deer Park (Niemi, 1973). The best time for hunting is after large storms when near-shore gravels are often remobilized and stony shoreline bluffs have been eroded. Readers wishing to try their luck are encouraged to begin at the Gitche Gumee Museum in

FIGURE 5.14. Omars form when spherical concretions are weathered from the surrounding rock matrix, leaving stones with smooth, spherical cavities. (Courtesy of National Park Service)

Grand Marais and are reminded that agate hunting is illegal within Pictured Rocks National Lakeshore.

Granite is one of the most common rocks found along beaches in the area and can be distinguished by its large crystals colored light gray, white, pink, or red, commonly peppered with about 15 to 20 percent dark-colored (black) minerals (Fig. 5.12b). Epidote is another common beach stone that exhibits an unusual black-green to yellow-green color (Fig. 5.12c).

Some of the more unusual beach stones are informally called omars, after the Omarolluk Formation of eastern Hudson Bay (Prest and others, 2000), an important source for these types of stones. Omars form as concretions (spherical accumulations of mineral material resembling ball bearings) are weathered from the surrounding rock matrix, resulting in a gray beach stone with distinctive smoothed spherical cavities (Fig. 5.14).

Large boulder-sized glacial erratics are another common feature found within the Pictured Rocks region. Erratics are pieces of rock that were eroded, transported, and then deposited by the glacier far from their original source area. Most erratics in the Pictured Rocks vicinity are of igneous or metamorphic origin and were

FIGURE 5.15. Boulder lag along the beach between Hurricane River Campground and Au Sable Point. The boulders were originally mixed in with the drift deposited here by the glaciers. Winnowing of the finer materials by wave action has left only the boulders behind. The boulders are mostly igneous and metamorphic rocks with source regions in Canada.

transported from locations on the Canadian Shield (the vast, low-relief terrain of central Canada). They are often found concentrated along the modern beach, where shore erosion has removed them from the surrounding drift, leaving the rocks as a significant shoreline "lag" (Fig. 5.15). This is best observed along the beach between Hurricane River and Au Sable Point (see road log). The largest erratic known to the author in the region is located on private land just east of the park boundary near Grand Marais. Called "the meteorite" by locals, it is in fact an angular-shaped, banded iron formation boulder of Precambrian age about the size of a large pick-up truck. Those interested in observing this specimen should inquire locally in Grand Marais.

THE HUMAN IMPRINT

Although human activity may seem far removed from the geologic story of Pictured Rocks, most geoscientists would agree that human activity is the premier geological agent operating in many late Holocene landscapes. Indeed, recent studies suggest that the magnitude of human actions in some landscapes equal or exceed more "natural" processes like stream or wind erosion. Thus, humans can significantly affect geological processes through changes in runoff and sediment transport, and these effects are important in creating modern landscapes.

Native Americans

Native Americans first arrived in the Great Lakes region approximately 12,000 years ago, following the margin of the melting glacier (Cleland, 1992; Lovis, 2009). In the northern Great Lakes, archaeologists divide the subsequent continuum of Native American cultural development into four episodes: the Paleo-Indian Period (12,000 to 8,500 years ago), the Archaic Period (8,500 to 1,800 years ago), the Woodland Period (1,800 to 360 years ago), and the Historic Period (360 years ago or 1640 A.D. to the present). Cultural resource surveys by the National Park Service and the U.S. Forest Service (Clark, 1993) have examined a large number of archaeological sites in the Pictured Rocks vicinity, many associated with ancient shorelines of Lake Nipissing. These sites are concentrated along the southern end of Grand Island (Popper, Duck Lake, and Trout Point sites), on or near the adjacent shoreline (Shelter Bay, Pedersen, and Big Bear sites), or within Pictured Rocks National Lakeshore (Miners Beach, Miners River, Mosquito Ridge, and Chapel area sites) (Anderton, 2002; Anderton, personal communication, 2010). Only the Trout Point site has been investigated using large-scale excavation (Benchley and others, 1988).

Archaeologists recognize at least 61 archaeological sites in the national lakeshore (Anderton and others, 2009; Legg and Anderton, 2010). Most are Woodland and Archaic Period seasonal habitation sites. They tend to be located near the mouth of streams, near paleolagoons on ancient barriers, in coves in sandstone bedrock along Lake Superior's shoreline, and along inland lakes.

The Miners Beach site is typical of many prehistoric sites in the Upper Great Lakes region, containing fire-cracked rock from hearth fires and an extensive scatter of quartz and quartzite flakes associated with stone tool making. During the Nipissing phase, the Miners Beach area was a small embayment rimmed by high sandstone bluffs along its southern, eastern, and western margins. Datable charcoal from hearths associated with fire-reddened sand exposed in the bluff face yielded calendar ages of approximately 3,370, 3,175, and 1,030 years ago. No copper tools or culturally distinct artifacts were found, but the dates indicate that the site is at least Late Archaic in age and was used into the Middle or Late Woodland Period (Anderton, 2002). Dunham and Anderton (1999) report a range of dates spanning the Late Archaic through Late Woodland Periods from the nearby Popper site on Grand Island, indicating that repeated, low-intensity use of shoreline sites was common throughout the region.

What was the purpose of these sites, and did they represent permanent habitation? Dunham and Anderton offer one possible explanation (1999, p. 16–17):

> It has long been held that Archaic peoples in the Upper Great Lakes followed a highly diffuse hunting and gathering strategy that involved the scheduled use of seasonally available food resources. . . . This pattern included seasonal mobility that placed people in coastal settings during the warm season and in interior settings in the cold season. Late Archaic peoples were the first to exploit fish in the region, although hunting appears to have been a more important activity. . . . Late Archaic fishing is thought to have focused on spring spawning species such as sturgeon and sucker that run in significant numbers in the streams and shallow Great Lakes coastal waters. . . . The Popper site has been interpreted to represent a coastal fishing camp and has provided evidence for multiple, possibly seasonal occupations during the Late Archaic period.

Thus the archaeological evidence indicates that by the time of European contact, Native Americans had been frequenting the Pictured Rocks area on a seasonal basis for thousands of years, although their exploitation of the region evolved over time, with fishing becoming much more important by the Late Woodland Period (A.D. 800). Interestingly, use of the park by Native Americans probably peaked during the Archaic Period, when better habitat and flooded river mouths provided maximum resources (Anderton, personal communication, 2010).

The Sawmill Culture

Geologically speaking, native peoples had little effect on the Pictured Rocks landscape. That dubious honor would be left to the generation of American lumbermen who began large-scale industrial clear-cutting operations in the region in 1876, and who, in just a few decades, cut much of the virgin forest between Munising and Grand Marais (Vogel, 2000).

The negative effects of nineteenth-century logging practices on Great Lakes ecosystems have been extensively documented by a number of scientists (e.g., see Ahlgren and Ahlgren, 1983). Beyond the profound ecological changes that result, clear-cutting generally leads to increased soil erosion and runoff, which leads to enhanced sediment loading of streams and lakes. Some logging techniques were especially damaging to riparian environments. Pines were the first to be cut, typically in the winter, and often transported to regional sawmills on snowmelt-swollen streams in the spring. Small dams were often constructed across the streams to enhance the discharge and hydrologic gradient. Once a sufficient pool of water was impounded, the dam was dynamited and the flooding water carried the logs downstream (Karamanski, 1989; Loope and Holman, 1991). Hardwoods were harvested after the pines were gone and were transported to the sawmill by rail, another practice with negative environmental consequences.

Loope (1993) studied the effects of a small logging dam on Beaver Lake and its natural outlet to Lake Superior, Beaver Creek. In 1905 a logging dam approximately five feet high was built near the mouth of Beaver Creek just upstream from Lake Superior, and was used for five to ten years. After its abandonment, the dam continued to catch logging and natural debris, creating a permanent obstruction to flow and causing annual high water extremes to back up each spring into both Beaver and Little Beaver Lakes. Employees of a small resort then operating on Beaver Lake, tired of their boathouse being flooded each spring, successfully removed the obstruction by the early 1960s. Today these lakes are ringed by sandy terraces approximately 1 1/2 feet above lake level that were periodically inundated in the spring floods. These processes had profound, although geographically limited, effects on the ecology and sedimentation history of Beaver and Little Beaver Lakes as well as the associated hydrology of Beaver Creek.

The environmental legacy of industrial-scale clear-cutting, however, was to prove much more insidious than the localized destruction of riparian habitat. By the time it was over, hundreds of thousands of acres had been completely wiped clean, burned over, and left in desolation. The 7,000-acre Kingston Plains, in particular, have become a national poster child for the rapaciousness of nineteenth-century lumbering practices (McKee, 1988; Fig. 5.16). Once home to a magnificent stand of virgin white pine (Frederick and others, 1976), these bleak stumped plains have resisted nearly 80 years of reforestation efforts and continue to be dominated by herbs, shrubs, grasses, and lichens (Lytle, 2005).

FIGURE 5.16. Kettled outwash of the Kingston Plains. These bleak stump plains have resisted nearly 80 years of reforestation efforts.

The persistent sterility of this environment has generated considerable scientific interest. Linda Barrett, a geographer now at the University of Akron, studied relationships among soils, vegetation, and fire on the Kingston Plains with her advisor, Randy Schaetzl of Michigan State University (Barrett, 1995, 1997; Barrett and Schaetzl, 1998). She compared soil profiles of nearby reforested areas with those of the Kingston Plains and found that the soils under reforested areas were slightly better developed and contained ortstein, a subsurface horizon (B horizon) cemented by iron and organic matter. The causal links between a lack of ortstein and the persistence of stump plains could not be explained, but she concluded that a number of factors might be involved in inhibiting reforestation, including prelogging forest vegetation and wildfire history. Unpublished data from nearby Hiawatha National Forest (Anderton, personal communication, 2002) suggest that the presence of thin clay lenses within the underlying outwash may have an influence on surface sterility. These "clay drapes" or "stringers" are common in braided outwash streams and are often deposited during waning meltwater floods. Because they are much less permeable than the overlying sands, the drapes are able to "perch" groundwater percolating downward through the soil, leading to slightly wetter and less drought-prone soil conditions. Preliminary evidence suggests that forests are often able to regenerate where the clay lenses are present, but where they are lacking, sterile stump barrens may result. Confirmation of this hypothesis awaits further study.

David Lytle of the U.S. Forest Service studied the pollen and charcoal record preserved in four kettle lakes on the Kingston Plains (Lytle, 2005). He showed that between about 9000 and 5000 years B.P., vegetation communities similar to the present stump barrens occurred at least three times and were closely linked to distinct periods of increased forest fire prevalence possibly driven by warmer, dryer climates. Although the factors involved are exceedingly complex, his data indicated that the Kingston Plains were especially vulnerable to natural disturbances during the early and middle Holocene and that human disturbance brought on by logging led to similar results in the late Holocene.

To many, the ecological carnage represented by the Kingston Plains is difficult to excuse, and few can visit the area without a sense of anger. It was Northern Michigan's misfortune to be exploited by a form of particularly egregious freewheeling capitalism that allowed the lumber barons to take whatever they wanted and leave the mess to others, as expressed by Russell McKee (1988, p. 68) in Audubon magazine:

> A prime example of the many rewards to be had at the time from unbridled lumbering practices and tax evasion lies in the career of General Russell Alexander Alger. Alger bought up a large portion of the Kingston pinelands and made a fortune in lumber. With his money he became governor of Michigan, then Secretary of War during the Spanish-American War, then a U.S. senator. When the timber was gone he dumped his holdings on the public, left the taxes unpaid, and went his way. Alger County, the land he helped wreck, bears his name.

SAND POINT: A LESSON IN GEOLOGIC COMPLEXITY

Sand Point provides an excellent example of how even the best-intentioned human modifications of natural systems can potentially lead to unexpected consequences. The point's location in protected waters near the entrance to Munising Bay (see Fig. 1.2a) made it a logical place for a Coast Guard station, which, after its decommission in 1960, was taken over by the park service. The main house now serves as the park headquarters, and the boathouse has been turned into a maritime museum. High lake levels during the mid-1980s led to a rapid retreat of the Sand Point shoreline, endangering (in the minds of some) the park headquarters building and flooding its basement. Alarmed park officials asked geographers Pat Farrell and John Hughes of Northern Michigan University to evaluate the situation. In their study, Farrell and Hughes (1984) used aerial photography taken since 1960 to analyze changes in the shoreline over time. They concluded that the permanently vegetated part of the point (as opposed to the sandy beaches at its periphery) had changed little over the period of record and that it appeared that Sand Point was in equilibrium with long-term fluctuations in lake level and sand supply. They also noted that Sand Point's shoreline had retreated in a similar fashion many times during the last century, and that the erosion was unlikely to threaten the park headquarters. Concerned park managers, charged with protecting the park's infrastructure, decided to err on the side of caution and built a 900-foot-long stone revetment on the north side of Sand Point to protect the headquarters (Fig. 5.17). By the time the revetment was in place, however, lake levels had dropped.

FIGURE 5.17. A rock revetment, built at Sand Point in the mid-1980s to protect the park headquarters from wave erosion, has had unexpected consequences. (Photo by Tony Williams, used with permission)

Over the next twenty years, lake levels never again reached the level of the mid-1980s, and the revetment never had an opportunity to provide protection from shoreline retreat. Indeed, the cycle of shoreline retreat reversed itself at approximately the position of the permanent vegetation line, just as Farrell and Hughes had predicted. Meanwhile, another more ominous set of problems began to emerge. Beyond the west end of the revetment, a new area of shoreline retreat, dubbed an "erosion hot spot," began to eat away at the Sand Point shoreline, concerning park managers (Fig. 5.17). Equally disturbing was the fact that despite historically low lake levels (which in the past had led to the construction of wide peripheral beaches), beaches remained narrow, and a large, new sandbar began to form far offshore on the north side of Sand Point. Never had such a bar been reported in the historical records, nor had it ever shown up in the aerial photographs. Concerned park officials asked Robert Young of Western Carolina University, an expert on coastal erosion, to evaluate the situation.

In his report, Young (2004) concluded that the rock revetment, by disrupting the natural longshore movement of sediment and waves in the area, was causing the "erosion hotspot," the narrow beaches, and the offshore sandbar. He provided the park service with three possible actions: (1) build another revetment along the southern shoreline, (2) remove the revetment completely, or (3) do nothing. Young recommended removal of the revetment as the most desirable option:

> It seems that reestablishment of natural processes is the best way to halt the current erosion, protect cultural resources, protect infrastructure, and meet NPS strategic goals. The buildings on Sand Point have outlasted many rises and falls in lake level. They have seen the shoreline grow and retreat numerous times. That infrastructure has not been threatened. The only thing that is different now is the existence of the rock revetment. Farrell and Hughes recommended against it. They were right. (Young, 2004, p. 8)

More recently, geologist Tim Fisher and his colleagues (Fisher, 2008; Castaneda and others, 2009; Craft and others, 2009) have been studying the Sand Point area and the series of bogs and beach ridges along its southeastern base, supported in part by research funding from the National Park Service. One of the goals of the research is to provide a mid- and late Holocene context for the more recent shoreline events. Preliminary data indicate that Sand Point postdates the Nipissing lake level and likely dates to the beginning of the Sault phase (about 1,500 to 2,200 years ago). The western half of Sand Point containing the headquarters infrastructure is younger than 700 years and perhaps as young as a few hundred years (Fisher, 2008).

As of 2011 the rock revetment was still in place. Because lake levels are falling, the revetment will be assessed again using grant funding from the Great Lakes Restoration Initiative to make sure nothing has been overlooked. If no objections are raised, it will likely be removed. While researchers are confident in the wisdom of the move, some park employees remain leery of removing what they see as their last bastion of defense against an unpredictable adversary. Regardless of the outcome, the situation at Sand Point is a valuable case study in the complexity (and uncertainties) of geologic systems with multiple variables, providing a real-world example of surface processes within a scale and context that are easily accessible. It also shows that geology is not just about ancient rocks and fossils, but also about people, places, and the processes operating in the modern landscape.

APPENDIX:
PICTURED ROCKS ROAD LOG

This road log is designed to take the interested reader to many of the park's best geologic sites. It begins in Munising and ends 84 miles later in Grand Marais, traversing the park from west to east following paved County Road H-58. A few of the side roads, such as the access road to the Chapel parking area, are gravel, but are well maintained. The log also includes a 10-mile loop hike to Chapel Beach and the Mosquito River mouth, two of the best places to observe bedrock geology in the park. The hike itself takes the better part of a day to complete. If this option is chosen, the suggested time to allow for the completion of the road log would be two days.

Names of the pertinent topographic quadrangle maps covering the route are provided for interested readers. Quadrangles are the standard topographic maps used by geologists throughout the United States. They are printed by the U.S. Geological Survey (U.S.G.S.) and are the starting point for nearly all geological investigations. Although several types have been produced, the 7.5 minute maps, which cover 7.5 minutes of latitude by 7.5 minutes of longitude (or about 50 square miles), have supplanted all others in common usage. These maps can be downloaded for free from the Michigan Department of Natural Resources website (www.Michigan.gov/dnr) or purchased at sporting goods stores. Readers may want to download or purchase these maps before attempting the road log. They will then be able to follow the road log (roads are shown on U.S.G.S. maps) on the quadrangles. Every time the road leaves one map and enters another, the name of the new map is inserted in the log. For those without maps, the road log route is shown on Figs. 1.2a and 1.2b.

Readers are reminded that car odometers vary slightly and that these variations, when extended over 84 miles, may produce readings slightly different than those given in the log. Periodic adjustments to the log mileage may be necessary.

MUNISING 7.5 MINUTE QUADRANGLE MAP

Munising

0.0 miles   Begin in the parking lot of the Pictured Rocks/Hiawatha National Forest Visitor Information Center near the corner of M-28 and County Road H-58 in Munising (Fig. 1.2a).

The village of Munising is built on a small cuspate lake plain that was formerly the bottom of Lake Nipissing. Between the base of the bluffs surrounding the southern edge of town and the modern shoreline, several beach ridges and terraces exist, recording the drop from the Nipissing level. The most prominent is a spit or barrier upon which State Highway 28 traverses the north side of the business district (Fig. 3.17). At least two small shoreline bluffs, possibly related to the Algoma lake phase, are observed in Bayshore Park next to the Pictured Rock Cruises boat dock (Fig. 3.18).

Turn right out of parking lot and proceed northeastward on H-58.

1.2 miles   Turn left onto Washington Street (just past sign for Sunset Motel and before hill) following signs for Munising Falls and Sand Point. Easy to miss!

1.75 miles   Washington Street becomes Sand Point Road, Memorial Hospital on left.

INDIAN TOWN 7.5 MINUTE QUADRANGLE MAP

1.8 miles   Turn right into parking lot of Munising Falls. A number of exhibits related to the geology of the region are on display here at the Visitor Center. Walk trail to the falls (800 feet).

Munising Falls

Munising Falls is one of the best places to observe the Au Train and Munising Formations up close (Fig. A.1), although the rocks are often obscured by lichens and moss. Here, Munising Creek has removed the overlying glacial debris and excavated an impressive valley into the underlying Cambro-Ordovician escarpment. The

Ordovician Au Train Formation is exposed as a distinct brown to white resistant cap rock at the top of the falls. Beneath it is the Upper Cambrian Miners Castle Member of the Munising Formation, which is weaker than the Au Train and has formed a deep alcove behind the waterfall. Visitors previously could walk behind the falls, but a recent collapse of the overlying bedrock onto the walkway forced its closure by park officials (those wishing to tempt fate in this manner will have a chance at the next stop).

FIGURE A.1. Munising Falls. The upper 3–6 feet of the falls is held up by the resistant Au Train Formation of Ordovician age. The rest of the falls, including the broad alcove behind the falls, is developed upon weaker strata of the Miners Castle Member of the Cambrian Munising Formation.

As described in chapter 2, the Munising Formation represents the flooding of the Laurentian landmass by shallow Paleozoic seas during a regionally widespread event known to geologists as the Sauk transgression. A close examination of the Miners Castle Member along the trail will reveal thin inclined surfaces (called cross-beds) within the beds made by currents along the ancient sea floor. Some adjacent beds lack cross-bedding because Cambrian organisms destroyed the original bedding as they sifted through the sands searching for food. These layers are said to be bioturbated, and they are common in the upper part of the Miners Castle Member. The best exposures here can be viewed from the right stairway.

Return to car. From parking lot, turn left on Sand Point Road / Washington Street. Optional: a right turn here will take you to Sand Point and the park headquarters, which is housed in an old Coast Guard station. A maritime exhibit is located nearby. On the way is a fine swimming beach on the left and, across from it, a pleasant loop trail through late Holocene bogs and beach ridges, some related to the Algoma lake phase and the formation of Sand Point, a well-developed spit. The road ends in a loop near the park headquarters. The stone revetment at Sand Point described in chapter 5 is located along the shoreline west of the loop (Fig. 5.17).

MUNISING 7.5 MINUTE QUADRANGLE MAP

2.4 miles   Road Ts at H-58. In front of you and slightly to the left is a metal stairway that marks the beginning of a foot trail to Olsen and Memorial Falls, which are similar to Munising Falls. The road log below will take you to these falls from a somewhat safer route with easier parking. Those wishing to walk the trail from the metal stairway should turn right onto H-58 and proceed about a 1/4 mile past the sign for the Sunset Motel, park at a small water pumping station on the right, and walk back to the stairway. Watch for speeding traffic!

Turn left onto H-58. Proceed up hill 0.3 miles.

2.7 miles   Turn right onto Nestor Street. Proceed 0.2 miles to small sign for Michigan Nature Association (MNA) trailhead and park on the right-hand side (1460 block of Nestor Street).

Memorial and Olsen Falls
Two impressive waterfalls, smaller than Munising Falls, can be viewed by walking the trail from this spot. The site is included in the road log because it is a good place to observe the Au Train and Munising Formations in a natural setting free from the prescriptive walkways and railings of Munising Falls. It also allows visitors to walk behind the falls (at their own risk, of course).

Trailhead signs farther down the path refer to the two waterfalls as M (Memorial) and O (Olsen). Olsen Falls was originally named Tannery Falls. The Michigan Nature Association, a nonprofit organization dedicated to preserving natural habitats in Michigan, manages the site. The trail descends gently into the woods from Nestor Street and crosses a small stream about 5 feet from the brink of Memorial Falls. Continue on the path around the edge of the cliff to a railed inclined path on the right, which takes you down to the falls and to excellent views of the Au Train and Munising Formations. This waterfall is not as impressive as Olsen Falls, but it is the easiest falls to walk behind. Laws of gravity here are strictly

FIGURE A.2. Olsen (Tannery) Falls. Although less impressive than Munising Falls, this feature allows better access to the rock faces and allows visitors to walk behind the falls.

enforced: be aware that collapse of the overhanging rock ceiling could occur at any time. Just to the left of the falls as you look at it, a tributary creek (usually dry) has cut another "twin" waterfall into the sandstone cliffs. This feature can be accessed by following a trail from the bottom of the falls up and along the left side of the valley to a natural window cut through the sandstone, which leads to the other falls.

Retrace the trail back to the top of the incline. Continue straight (away from the car) on the main trail from Nestor Street (following yellow arrows), which rises and then eventually descends steeply through the woods down a long hill. Highway 58 (H-58) is at the base of this hill. Just before reaching the road (you should be able to hear the traffic), the trail to Olsen Falls leads off to the left and then up a small hill. Following this trail another 1/4 mile or so brings you to Olsen Falls (Fig. A.2).

Olsen Falls is the more impressive of the two falls owing to the greater discharge of Tannery Creek. The back of the falls is less accessible than at Memorial Falls but the alcove-forming aspect is especially evident because the stream often flows back along the base of the alcove.

Return to car at Nestor Street. Turn car around and retrace your route back down Nestor Street to H-58.

3.1 miles   Junction H-58. Turn right onto H-58. Continue on this road for the next 3.7 miles.

INDIAN TOWN 7.5 MINUTE QUADRANGLE MAP

4.7 miles   At about this point, continuing east on highway H-58, we begin to enter a region of thin drift and glacially fluted terrain with drumlins. Ice movement was from left to right across the road. Drumlins are not visible from the road. The flutes are subtle and difficult to pick out from a moving car, but are conspicuous on topographic maps and aerial photographs (Fig. 1.2a).

5.15 miles   Road rises onto flutes. Golf course on left.

5.5 miles   Road descends flutes.

5.8 miles   Road rises onto low-relief fluted upland.

6.1 miles   Just past Indian Town Road on right, road crosses small creek and rises onto another small flute.

6.8 miles   Turn left (north) onto Miners Castle Road (County Road H-11). Continue on this road for next 5.3 miles to Miners Castle parking lot.

This road continues across the area of thin drift with northwest–southeast oriented flutes. Ice movement was from left front to right rear.

9.7 miles   Road descends into a small meltwater channel carved into bedrock at 827 feet altitude. This channel drained the melting glacier as it retreated from this position approximately 11,400 years ago. Drainage was from left to right.

10.5 miles   Road ascends the northern side of the 827-foot meltwater channel. Miners Falls road to the right. Stay on H-11.

11.8 miles   Junction, Miners Beach Road. Bear left continuing on H-11 to Miners Castle parking lot.

Miners Castle Area

12.1 miles   Enter Miners Castle parking lot. Park car and walk to Miners Castle overlook.

This site is one of the most well developed in the park and contains a small bookstore, modern restrooms, and extensive signage addressing the park's bedrock geology. The area contains two principal overlooks: an upper platform perched on the cliff's edge above Miners Castle (Fig. A.3), and a lower observation deck positioned adjacent to Miners Castle and accessible by a short path that descends to the east.

Miners Castle
Member

Chapel Rock
Member

FIGURE A.3. Miners Castle, the type locality for the Miners Castle Member, as seen from the Miners Castle overlook. The small cave just above water level is developed in the underlying Chapel Rock Member. Arrow shows the contact between the two members.

Miners Castle is the type locality for the Miners Castle Member of the Munising Formation (described in chapter 2). It was easily accessible in the 1950s when Hamblin designated it as the type section. With development of the park, however, Miners Castle can only be observed today from behind a railing at the lower observation deck, although limited areas for examining the rocks are available nearby. From the upper platform, a small portion of the underlying Chapel Rock Member is exposed in the cliffs just above the waterline and contains a small wave-cut cave (Fig. A.3). This cave continues beneath Miners Castle to an opening facing the open lake (Fig. 2.18) (Anderton, 2002). During storms, waves will move forcefully through the cave and discharge foaming water at the southern end. The remainder of the cliff is developed in the Miners Castle Member. The Jacobsville Formation is just below the waterline here.

Some controversy exists as to the nature of the Miners Castle landform. Dorr and Eschman (1970, p. 208) considered it a relict stack, formed by "an earlier and higher glacial lake." We now know that any such lake would have to postdate the Marquette ice advance, and no lake at the appropriate elevation has been recognized in the region. Nor is it likely that Miners Castle is an active stack formed by modern Lake Superior. Stacks typically form by the collapse of an arch into the water body below, leaving a stack or tower of rock stranded from shore. The turret of Miners Castle does not extend downward to the lake. A more plausible (though less exciting) explanation is that this feature is simply a product of differential weathering and erosion (Anderton, 2002, p. 44). As discussed in the previous chapter, the eastern (and larger) turret of Miners Castle collapsed into Lake Superior in 2006, rendering the landmark somewhat less impressive than before (Fig. 5.10).

Return to parking lot and head south on Miners Castle Road.

12.65 miles   Turn left on Miners Beach Road. Proceed 1.0 mile to T in road (right takes you to east end of beach, left to west end).

13.65 miles   Turn right and proceed 0.25 miles following signs for the "Lakeshore Trail" to parking lot at east end of Miners Beach.

Miners Beach (east)

Take the Lakeshore Trail path from parking area north toward Miners Castle about 0.25 miles to the beach.

This area contains two geologic sites. The first, referred to as the "east end" of Miners Beach, occurs where the sands of Miners Beach meet bedrock at the east end of the beach. This is the best and most accessible place to view mudstones at the top of the Chapel Rock Member, which are located just a few hundred feet to the northeast of a small waterfall at the beach's eastern margin. According to Haddox (1982, p. 74 and 77):

> The thickest mudstone lens [here] is 30 m wide and 1.4 m thick, and recedes into a bluff topped by slightly more resistant sandstone that appears to be transitional between the Chapel Rock and Miner's Castle Members. . . . Some thinly-interlaminated mudstone and sandstone are slightly mixed, apparently by tunneling organisms, to present nearly circular cross-sections of sand burrow-fillings millimeters in diameter and rimmed with wisps of silt. Other trace fossils (Rusophycus, Cruziana, and Planolites) are visible on bedding plane surfaces in this mudstone.

The second geologic site in the area, known as the "ledge" at the eastern end of Miners Beach (Hamblin, 1958; Haddox, 1982, Haddox and Dott, 1990), is located about 1/3- to 1/2-mile distant from the beach, along the rocky headlands to the north and east (the intervening coast and headlands are located on the adjoining Wood Island Southeast 7.5 minute quadrangle map). The site can be accessed by following a muddy, sometimes faint trail that follows the coastline out to the headland. The trail begins at the east end of Miners Beach, near a large concrete slab about 25 feet south of the small waterfall mentioned earlier. The path is difficult walking but the geology at the end is worth the walk. The promontory at the "ledge" affords an excellent view along the Pictured Rocks cliffs to the east. This site is one of the best places in the park to view large-scale high-angle cross-sets (possibly representing ancient subaqueous dunes) and reactivation surfaces (places where newer dunes cut across older ones) within the Chapel Rock Member (Fig. 2.12b) (Haddox, 1982). The surface of the rock ledge here truncates the cross-sets in places, so that they are exposed on edge in dramatic fashion (Fig. A.4). The contact between the Chapel

FIGURE A.4. On-edge view along the strike of thin cross-laminae to medium cross-beds exposed at the east ledge of Miners Beach. Tape measure is extended 10 centimeters (3.9 inches).

Rock and Miners Castle Members is also readily observable where differential weathering has formed a distinct bench lying about 3 feet above modern lake level (Fig. 2.8).

Return to car. Proceed back down the road you came.

14.15 miles   Stop sign. Continue straight ahead to west end of Miners Beach.

Miners Beach (west)

14.35 miles   Arrive at West End Miners Beach parking lot. Park car.

A sign near the bathrooms here directs visitors to a short trail to the beach. Take this trail until it intersects the Lakeshore Trail just before the main stairway to the beach. Follow the Lakeshore Trail left (or west) until it reaches Miners Creek. Cross the footbridge over Miners Creek. Immediately after crossing the bridge, turn right off of the Lakeshore Trail onto a trail leading to the beach along the base of cliffs on the west side of stream. The trail fades near the beach, but a bit of bushwhacking to the left (west) through pine thickets leads you to other faint trails that emerge at the beach. An alternative route is to descend the stairs to Miners Beach. Then walk west, fording Miners Creek (rarely greater than knee-deep) to reach the west end of the beach.

This area is known as the "west end" or "west" Miners Beach (Fig. 1.7) (Haddox, 1982). The Chapel Rock Member is exposed low in the section near modern lake level, with the Miners Castle above. Haddox and Dott (1990) studied large low-angle cross-sets in the Chapel Rock Member here (Fig. A.5), concluding that these strata likely represented "eolian, medium-to-coarse sandstones . . . imply[ing] dunes more than 5 m high and wind speeds in excess of 40 m/s (75 mph)" (p. 707). In contrast, horizontally bedded dark mudstones intermingled with sandstones and fine siltstones (indicative of slack-water conditions) are exposed at the top of the Chapel Rock Member nearby (Fig. 2.14). These contrasting styles of deposition within the Munising Formation, representing variable fluvial, tidal, shallow marine, and eolian conditions, have made interpreting these rocks especially challenging (Haddox and Dott, 1990).

FIGURE A.5. Low-angle cross-sets in the Chapel Rock Member, probably representing eolian (wind deposited) dunes (Haddox and Dott, 1990), west end Miners Beach. Tape measure is extended 1 meter (3.3 feet).

As discussed in chapter 5, Native Americans have frequented this area since at least the Late Archaic. During the Nipissing lake phase, the Miners Beach area formed a small cove in the Lake Nipissing shoreline. The low-relief flats behind the present beach are developed upon the former Lake Nipissing lake bottom. A series of east–west trending beach ridges located between the Nipissing and present shoreline record the transition to lower lake levels after the Nipissing high stage.

Return to parking lot. Retrace route back along road.

14.7 miles   Yield sign. Turn right onto Miners Beach road and return to junction with Miners Castle Road.

15.7 miles   Turn left (south) onto Miners Castle Road.

17.0 miles   Junction Miners Falls Road. Stay on Miners Castle Road. Optional: a left turn here will take you to the parking area for Miners Falls (Fig. A.6), developed on the Miners Castle Member of the Munising Formation. Trail to falls is approximately 1.2 miles round-trip.

20.7 miles   Junction H-58. Turn left onto H-58. For the next 8.4 miles the road continues to traverse fluted terrain with thin drift.

21.75 miles   Descend into bedrock valley of Miners River. The size, pattern, and orientation of this valley may be very similar to the postulated buried valley that formed the Kingston Lake kettle chain (Blewett and Rieck, 1987).

MELSTRAND 7.5 MINUTE QUADRANGLE MAP

24.7 miles   Junction H-15. Village of Van Meer. Continue straight ahead on H-58.

29.15 miles   Pass through village of Melstrand. The Melstrand store is the last commercial establishment until Grand Marais, 55 miles ahead via the road log. Melstrand marks the transition zone between northwest–southeast fluted terrain and the broad outwash plains

FIGURE A.6. Miners Falls is developed upon the Miners Castle Member of the Munising Formation. (Courtesy of National Park Service)

and related ice-wastage features containing thicker drift of the Kingston Plains.

29.45 miles   Just past Melstrand, turn left on Chapel Road toward Chapel-Mosquito area.

29.5 miles   Immediately after turning, pull off on the two-track on the right side of road and park. Cross to the other side of the road and walk approximately 200 feet north along Chapel Road (away from H-58). Turn left into the woods and walk approximately 200 yards to examine esker.

The ridge in front of you is a small part of a larger and more extensive northwest–southeast oriented sinuous ridge approximately 1/2 mile long. Because it is located in a zone where eskers, outwash plains, and crevasse fillings are superimposed upon fluted terrain and drumlins, some ambiguity exists as to the true nature of the landform. Park service personnel (Gregg Bruff, personal communication, 2008) informally refer to the feature as a drumlin. Well-developed drumlins in the area, however, are much larger, distinctly elliptical, and nonsinuous. The delicate sinuosity of the feature on topographic maps, its location within a zone of well-developed eskers of similar size, shape, and orientation (Melstrand Esker), and the presence of a shallow kettle (or abandoned gravel pit?) on its northwest side, all suggest that the landform is an esker. No extensive sediment exposures are available to confirm its origin. The surface is extremely hummocky, possibly from tree throw. The feature is best observed in the late fall, winter, and early spring when leaves are off.

Return to car and continue north on Chapel Road.

30.5 miles   Cabins on the right side of the road generally mark the location of the ice margin during the building of the Melstrand esker and associated outwash fan (Figs. 1.2a, 4.8), which are best developed in the woods to your right, about 1.0 mile to the northeast (directions to the esker are provided later in the road log). The cabins are built on the western fringes of the fan, near the ice-contact slope.

31.7 miles   Bear right, following sign to Chapel Area.

GRAND PORTAL POINT 7.5 MINUTE QUADRANGLE MAP

34.6 miles   Chapel parking area. Rocks of the Au Train Formation ring the parking lot. The trailhead here offers access to two of the most significant geological sites in Pictured Rocks National Lakeshore: Mosquito Beach and Chapel Beach. Both can be visited by means of a loop trail (9.7 miles or 10.4 miles depending on the exact route chosen, Fig. A.7) that takes the better part of a day to complete.

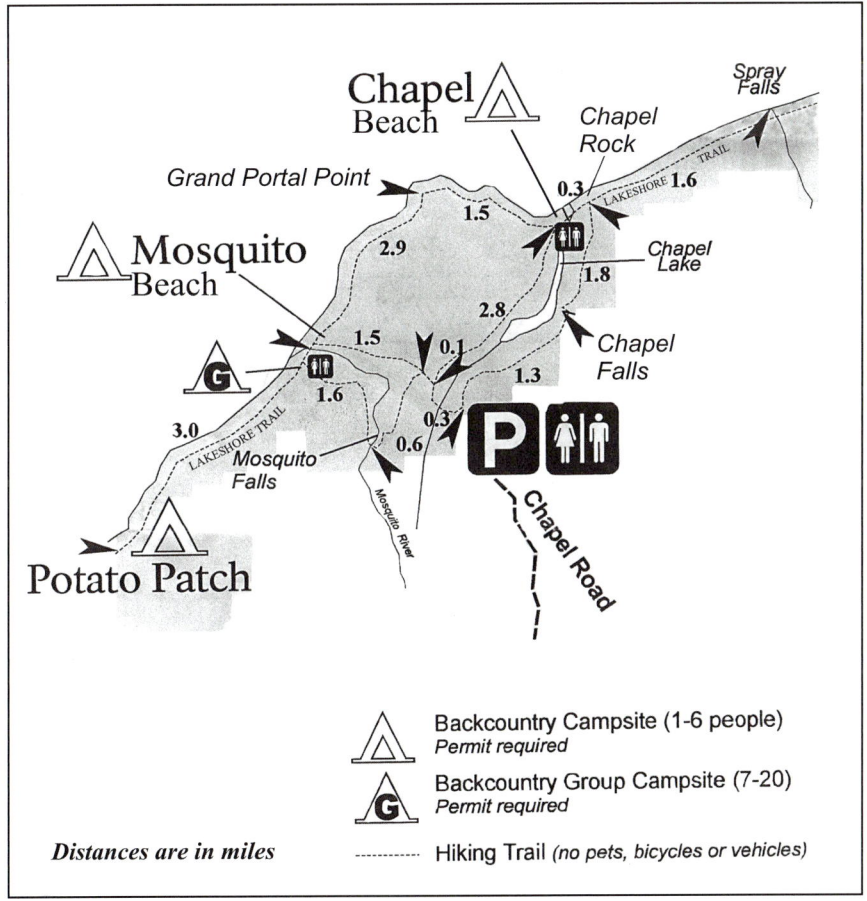

FIGURE A.7. Chapel Area trail map. (Courtesy of National Park Service)

The trail log that follows takes a 10.4-mile counterclockwise loop from the parking lot to Chapel Beach and Mosquito Beach via Chapel Falls (outbound) and Mosquito Falls (return). The mileage described here comes from the flyer "Chapel Basin Day Hikes"

(Fig. A.7) printed by the park service. Visitors can reduce the total mileage slightly by skipping Mosquito Falls on the return trip and taking an alternative route along the north side of Mosquito River (9.7 miles total). Although the trails are not overly strenuous, several steep or muddy sections do exist. Those not in excellent physical shape should consider visiting the Chapel and Mosquito sites on separate days. If time is limited, the Mosquito Beach site is closest and exhibits the most interesting and varied geology of the two.

Parking lot to Chapel Rock (0.0 to 3.1 miles on trail)—Take the trail to Chapel Beach via Chapel Falls. About 1/2 mile from the parking lot, the trail climbs several small meltwater terraces onto a bedrock upland, which it then follows along the southeastern side of Chapel Lake. Chapel Lake lies in the bottom of a deeply incised bedrock gorge that carried meltwater eastward along the edge of the retreating glacier approximately 11,000 years ago (Fig. 4.8f). At 1.3 miles, the trail reaches the Chapel Falls Overlook. The top of Chapel Falls is developed upon the Miners Castle Member of the Munising Formation. The Au Train Formation, the typical cap rock in the region, is exposed about 150 feet upstream, capping another small falls just below the footbridge. The trail descends the Pictured Rocks escarpment to another viewpoint of Chapel Falls, before continuing on to Chapel Rock. Just before reaching the Lake Superior shoreline, the trail descends a steep Lake Nipissing shore bluff and dumps the hiker at Chapel Rock. The trail Ts at the Lakeshore Trail here. After examining Chapel Rock, follow the Lakeshore Trail west (left) toward Chapel Beach.

Chapel Beach Area

Chapel Beach is the home of Chapel Rock, a natural sandstone portico (Figs. A.8, 2.11) defined as the type locality for the 30- to 65-foot-thick Chapel Rock Member of the Munising Formation (Hamblin, 1958). Exposures of the Chapel Rock Member here are confined to Chapel Rock itself, the bed of Chapel Creek, and cliffs at the west end of the beach. Most of the rest of the section is exposed in the adjacent headlands, which are best studied from the lakeside. Low-angle cross-stratification is exposed in the columns

FIGURE A.8. Chapel Rock, the type locality for the Chapel Rock Member of the Munising Formation, from Chapel Beach.

of Chapel Rock (Fig. 2.13), and a distinctive conglomerate facies can be observed at the base of the rock near the shoreline. High-angle trough cross-strata are exposed in cliffs along the west side of the beach. Interpreting low- and high-angle cross-strata in the Chapel Rock Member has been challenging and variously attributed to eolian, marine, and fluvial influences associated with a sandy tidal inlet or sandy delta (Haddox and Dott, 1990).

During the Nipissing phase about 5,000 years ago, the Chapel area was part of a small embayment in the coastline of Lake Nipissing, similar to Miners Beach. The campground and pine flats south of the modern beach are developed upon the old Nipissing lake bottom. Dorr and Eschman (1970, p. 97) interpreted the alcoves in Chapel Rock as sea caves cut by the waves of an ancient higher-level lake, presumably Lake Nipissing. Differential weathering and erosion during the late Holocene have also played a significant role in shaping the landform. Just west of Chapel Rock, Chapel Creek enters Lake Superior in dramatic fashion as a small waterfall.

Chapel Rock to Mosquito River (3.1–7.8 miles on trail)—Continue on the Lakeshore Trail to the west end of Chapel Beach, following the signs for Mosquito River. After leaving the beach, the trail climbs steadily to the top of the Pictured Rocks escarpment, passing the Battleships, a set of deep embayments and linear headlands controlled by joint sets within the Chapel Rock Member (Fig. A.9). Over the next 4 1/2 miles the trail follows the crest of the escarpment, in places nearly 200 feet above the surface of Lake Superior, reaching its highest point at Grand Portal Point. The views along this section of the trail are the most spectacular in the park—a backpacker's equivalent to California's coastal highway.

FIGURE A.9. Weathering and erosion along closely spaced joint sets have produced the distinctive Battleship Rocks located just west of Chapel Beach.

Mosquito Beach Area

Here, the Mosquito River tumbles to meet Lake Superior through a broad cleft in an otherwise precipitous coastline, forming one of the park's more picturesque natural settings. The river separates the site into an eastern and western half. The upper half of the Chapel Rock and lowest portion of the Miners Castle Members are exposed along the adjacent beach, with excellent displays of ripple marks, mud cracks, trough cross-bedding, and conglomerate facies within the Chapel Rock Member.

The eastern end of the Mosquito River area is the best place to observe ripple marks (Figs. A.10, 2.17a, b). Here, horizontally stratified to very gently dipping sandstone facies of the Chapel Rock Member occur in tabular sets separated by low-angle truncations that extend laterally for hundreds of feet, forming a broad horizontal sandstone bench. Many of these strata contain ripple marks, both of the oscillation and current variety. Linear grooves or tracks, possibly made by organisms, extend across the ripples (Fig. 2.17b). According to Haddox and Dott (1990, p. 705):

> At Mosquito Harbor, well-exposed symmetrical and asymmetrical ripple crests show a prominent northeast–southwest mean orientation with asymmetrical examples steeper toward the southeast. These orientations indicate that waves and possible flood-tidal currents approached from the northwest and were shoaling toward the southeast. The cross bedding orientation, however, demonstrates coarse-sediment transport toward the northwest for both the Basal Conglomerate and Chapel Rock sandstones, which is perpendicular to the shoreline trend inferred from the ripples.

In other words, at the time these sandstones were emplaced, sediments from the Northern Michigan highlands were being swept northwestward into the transgressing Paleozoic sea, along a coastline experiencing (at least some of the time) waves and currents from the northwest.

FIGURE A.10. Ripple marks with superimposed mud cracks in the Chapel Rock sandstone at Mosquito Beach. Tape measure is extended 10 centimeters (3.9 inches).

The eastern end of the beach is dominated by a conspicuous low scarp developed upon a fracture trending generally parallel to the beachfront (Fig. A.11). Farther east along the beach, well-developed truncated trough cross-sets are observed near the top of the Chapel Rock Member.

The west side of the beach can be accessed by crossing a small footbridge just upstream from the mouth of Mosquito River. This area marks one of the best places to view the channelized conglomerate and breccia facies of the Chapel Rock Member,

FIGURE A.11. Prominent fracture trace in the Chapel Rock sandstone at the east end of Mosquito Beach.

FIGURE A.12. Trough cross-beds in the Chapel Rock Member at the west end of Mosquito Beach. View looking north–northwest. Tape measure is extended 1 meter (3.3 feet) and oriented parallel to the direction of flow (into the photo).

which are exposed in the low cliffs along the back of the beach. These channels were likely cut and then rapidly filled by vigorous surface steams, providing further evidence for wide variations in flow intensities within the Chapel Rock Member.

Near the swash zone, northwestward plunging trough cross-beds are well exposed (Fig. A.12). Farther west, polygonal mud cracks are especially conspicuous and associated with the laminated mudstone facies of Haddox and Dott (1990) (Figs. 2.15, 2.16). Hamblin (1958) notes that these units are especially common near the top of the Chapel Rock Member and in many places are in direct contact with basal units of the Miners Castle Member. He notes that most of the cracks have been filled with sand and attributes their formation to shallow water environments that were repeatedly exposed to subaerial (exposed to the open air) conditions. Trilobite tracks can also be observed here by careful investigations along mudstone bedding planes.

Mosquito Beach to Chapel parking area via Mosquito Falls (7.8–10.4 trail miles)—Hikers have two options in returning to the Chapel parking area from Mosquito Beach. The shortest (1.9 miles) is along the north side of the river. The longer (2.6 miles) and more scenic trail via Mosquito Falls begins on the south side of the river. This trail actually crosses two waterfalls, the first associated with a narrow tributary canyon of Mosquito River. The second waterfall is Mosquito Falls.

Return to car. Proceed south on Chapel Road to H-58.

MELSTRAND 7.5 MINUTE QUADRANGLE MAP

39.7 miles   H-58 junction. Turn left.

40.0 miles   Road continues to climb onto the Melstrand outwash plain, a small fan fed by the Melstrand esker (Figs. 1.2a, 4.8).

41.0 miles   A two-track road leads back to the Melstrand esker on left. Continue straight on H-58.

Melstrand Esker
Optional: to view the esker(best when leaves are off), turn left onto this road. After about .6 miles, the two-track turns left and continues another .2 miles, before turning right and descending the ice-contact slope. Where the road turns right, a large fan perched atop the crest of the Melstrand outwash apron looms in the woods in front of you. This marks the downstream end of the esker. Park in the faded two-track on the left just before the road descends the ice-contact slope. Walk out the faded two-track to the northwest to view the esker along the left side of the road. This conspicuous esker is indicative of well-developed subglacial or englacial drainage systems commonly associated with warm-based or temperate glaciers in which surface meltwater can penetrate to the glacier bed. The esker merges downstream into the ice-contact slope of the Melstrand fan, indicating that subglacial meltwater streams were an important sediment source for many of the smaller outwash fans in the region.

CUSINO 7.5 MINUTE QUADRANGLE MAP

(road just catches the northwest corner of the map)

TRAPPERS LAKE 7.5 MINUTE QUADRANGLE MAP

44.7 miles   Road descends into Long Lake Channel.

Long Lake Channel

The Long Lake Channel (chapter 4, Fig. 4.8c) formed as meltwater trapped between the ice margin, and the Lower Kingston ice-contact slope south of Beaver Lake spilled southward. It exhibits an extensive 919-foot surface, with at least two smaller incised channels at 909 feet and 899 feet. Drainage was from left to right (south). A small overgrown gravel pit on the channel bottom near here contains a distinct gravel lag with rounded boulders up to 1 foot in diameter. Discharge may have been as large as 45 times the mean annual discharge of the modern Tahquamenon River.

44.8 miles   Road to Little Beaver Lake on left (optional; 6 miles round trip). Otherwise stay on H-58.

If taking the Little Beaver Lake Road: About 1 mile after the turnoff, the road descends the Cambro-Ordovician escarpment that forms the Pictured Rocks cliffs farther west. Here the escarpment has turned away from the lake and is covered in most places by thin drift. The road traverses several meltwater channels cut into bedrock (drainage was from left to right), before descending the Lake Nipissing scarp into Little Beaver Lake Campground. A trail leads from the parking lot just south of the campground to Twelvemile Beach and Lake Superior, a round-trip distance of 3.2 miles. This trail is an excellent way to experience the solitude of Twelvemile Beach in an area inaccessible by automobile. Both Beaver and Little Beaver Lakes were former embayments in Lake Nipissing. Sediment cores taken from Beaver Lake indicate that this area was ice free by about 10,700 years ago (Fisher and Whitman, 1999). Visitors can observe Little Beaver Lake from the campground. Beaver Lake is only viewable by foot trail.

45.5 miles   Continuing on H-58, the unpaved road on the left takes you to the Beaver Basin overlook (optional). As of 2011, road conditions were marginal for 2 x 4 vehicles with low clearance.

46.1 miles   Road climbs up out of Long Lake channel and onto the Upper Kingston outwash plain (Fig. 1.2b).

47.6 miles   Junction with Twin Lake Truck Trail. Continue straight on paved H-58.

49.4 miles   Junction with Ross Lake Road. Stay straight on paved H-58.

AU SABLE POINT SOUTHWEST 7.5 MINUTE QUADRANGLE MAP

51.9 miles   H-58 and pavement turn wide to the left. Turn right onto Adams Trail, which is paved for a short distance. Continue straight on Adams Trail. Pavement ends. Proceed 2.0 miles.

Kingston Plains

As late as 1875, this area was a 5,000-year-old virgin white pine forest (Frederick and others, 1976). Within just a few decades the area was clear-cut, burned (perhaps multiple times), and abandoned for back taxes, leaving a ravaged stump plain of lichens and small brush. Unlike surrounding areas, the dozen or so square miles here resisted reforestation, although the exact mechanisms are a matter of debate (see chapter 5). Indeed, only in the last 30 years or so have mature trees begun to take hold. Kettled areas in particular seem to defy reforestation and present an interesting problem for scientists. The area continues to serve as a symbol for the rapaciousness of late nineteenth-century logging.

53.2 miles   Junction with gravel road (Nugent Lake Lane). Stay straight on Adams Trail.

53.8 miles   Small parking area on right for those not wishing to park on road (see next item).

53.9 miles Pull off on right side of road (watch out for deep sand). Leave car and cross to the other side of the road. Proceed to the lowest spot on this stretch of road and examine kettles on both sides of road.

Kingston Lake Kettle Chain

Several stumped kettles are visible in the distance from this location on both sides of the road (Fig. 5.16). They are part of the 10-mile-long, northwest–southeast trending chain of dry kettles and kettle lakes known as the Kingston Lake Chain (see chapter 4). The main chain is echoed by two additional chains that converge to the north in a pattern resembling a dendritic stream system (Figs. 1.2b, 4.7a, b). Nearby Nugent Lake is located in one of the tributary chains. Blewett and Rieck (1987) attributed formation of the chain to a buried valley in the preglacial surface, possibly formed in bedrock, and similar to the modern Miners River valley. Stagnant ice lingered in this valley during deglaciation and was covered by outwash. The ice blocks later melted out, forming the kettle chain (Fig. 4.9). At least 20 dry kettles and kettle lakes can be viewed in a short walk along the axis of chain from here toward the northwest (Fig. A.13).

FIGURE A.13. Well-developed dry kettle in the Kingston Lake kettle chain. Note person for scale (arrow).

Return to car. Carefully perform a U-turn and proceed back to H-58 intersection.

55.9 miles   Junction with H-58. Turn right, following sign for Kingston Lake Campground. Proceed north across Kingston Plains. You are climbing the Kingston outwash plain, a large head of outwash associated with a major ice-marginal position. The ice margin was in front of you and meltwater drainage was toward you.

59.25 miles   Begin descending the ice-contact slope of Upper Kingston outwash apron. This hill marked the approximate position of the ice margin approximately 11,400 years ago (Figs. 1.2b, 4.7, 4.8). Kingston Lake on left.

60.5 miles   Turn left into Kingston Lake Campground. Take first left into the Boat Launch parking lot. Park and view lake.

Kingston Lake is the largest lake in the Kingston Lake kettle chain (Fig. A.14). Hughes (1968) named the surrounding outwash surface the Lower Kingston surface (Surface B on Fig. 4.8c).

Return to car and return to H-58.

FIGURE A.14. Kingston Lake, the largest lake in the Kingston Lake kettle chain.

60.7 miles   Turn left onto H-58.

61.05 miles   Climb out of the Kingston Lake depression and on to the Lower
Kingston surface (Surface B on Fig. 4.8c). Outwash was flowing
toward you and from left to right, controlled by a high bedrock sill
just south of Grand Marais. From here all the way to Grand Marais
(a distance of 23 miles) the road follows a stair-step-like series of
kame terraces, progressively lower toward the north, truncated
near the lakeshore by the Nipissing beach scarp. The terraces are
the kame terraces described in chapters 3 and 4 and shown in
Figs. 4.8a–f. The nomenclature from this figure will be used in
designating these terraces for the purpose of the road log.

From Kingston Lake to Twelvemile Beach Campground the road
descends these terraces all the way down to the lake, where they are
truncated by the Nipissing bluff (Fig. 4.7a). Between Twelvemile
Beach and the Log Slide, the road climbs back up the terrace
stairway, and between the Log Slide and Grand Marais, the road
performs a complicated dance up and down the terraces, as it
works its way around the southern edge of the Grand Sable Dunes.

62.0 miles   Junction with Hurricane River Truck Trail on right. Continue (bear
left) on paved H-58.

62.1 miles   Enter collapsed outwash associated with the Lower Kingston
outwash plain (Fig. 4.7a; Surface B, Fig. 4.8). This area was very
close to the ice margin at the time that Surface B was forming.
Outwash was deposited on top of stagnant ice, which later melted
to produce the hummocky topography observed. Northern
extensions of the Kingston Lake kettle chain also pass through
this area.

62.7 miles   Optional: Turn left into two-track and park. Due west of this location
(SE 1/4, Sec. 30, T.49 N., R. 15 W.), approximately 1/3 mile from the
road, sits a 130-foot-deep kettle lake associated with the Kingston
Lake kettle chain named Turtle Lake (Fig. A.15). This may well be the
deepest kettle in Michigan, although solutional enhancement of the
subcropping dolomitic Au Train Formation cannot be ruled out.

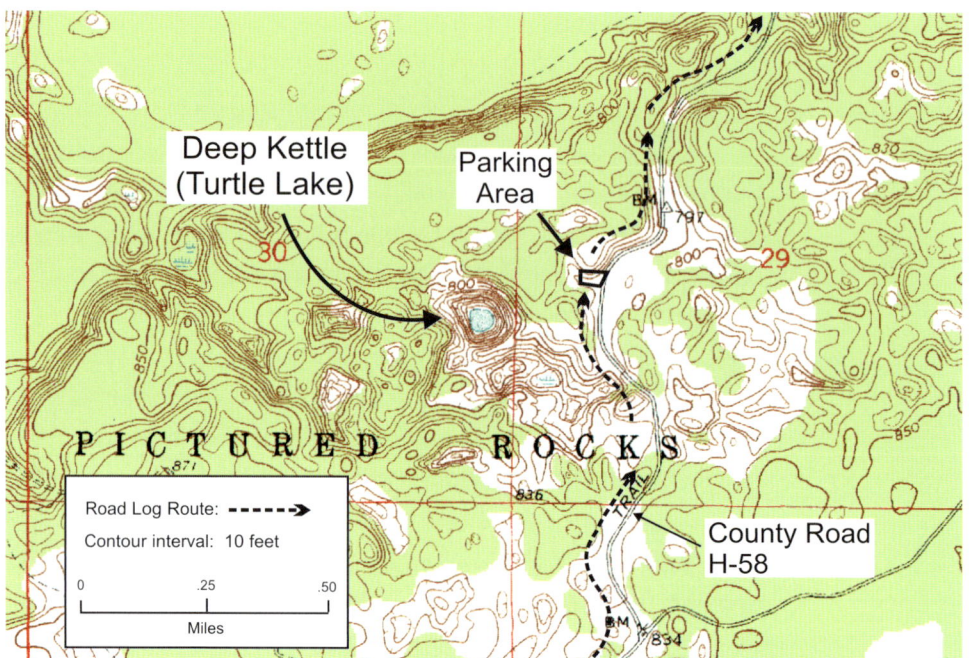

FIGURE A.15. Portion of the Au Sable Point Southwest quadrangle showing a deep kettle lake informally named Turtle Lake.

63.1 miles   Begin long descent of the Lower Kingston ice-marginal position (Figs. 4.7a, b, 4.8c).

63.8 miles   Emerge onto the 730-foot (225 meters on diagram) kame terrace, the highest of the Beaver Basin surfaces (Figs. 4.7a, 4.8d). Meltwater was traveling from left to right.

AU SABLE POINT 7.5 MINUTE QUADRANGLE MAP

64.5 miles   Turn left onto road to Twelvemile Beach Campground.

65.15 miles   Road descending Nipissing beach scarp. Go past road to campground on left and continue straight ahead to picnic area parking lot.

65.2 miles   Parking lot. Park car and walk down stairs to view Twelvemile Beach, one of the loneliest stretches of Lake Superior coastline in Michigan. The coastline is lower here because the Cambro-Ordovician escarpment that makes up the Pictured Rocks cliffs farther west has turned inland and is now several miles to the south.

Return to car and drive back to H-58 intersection.

66.0 miles H-58 junction. Turn left toward Grand Marais.

66.6 miles Road begins descent of Nipissing scarp and beach ridge complex, which ends at approximately 67.2 miles.

68.1 miles Turn left into Lake Superior overlook parking lot. This site contains a viewing platform overlooking Lake Superior and the mouth of Sullivan Creek, plus a stairway to the beach.

Return to car. Turn left onto H-58.

68.8 miles Road ascends Nipissing beach scarp.

69.25 miles Road descends Nipissing beach scarp just before turn off to Hurricane River Campground.

69.3 miles Junction, road to Lower Hurricane River Campground. Turn left onto this road (Lower Hurricane River Road) just past the stone bridge. Go past the "day use" parking area on left near the corner.

69.4 miles Proceed to parking area. Walk to mouth of Hurricane River.

Hurricane River, Au Sable Point, and the Jacobsville Formation
The stretch of shoreline between Hurricane River and Au Sable Point contains the park's best exposures of the Jacobsville Formation, the oldest formation exposed in the park. The Hurricane River parking lot is the most convenient access point to the rocks and to Au Sable Point, home of the picturesque Au Sable Point Lighthouse (Fig. A.16).

Near the parking lot, the Jacobsville Formation is exposed in a small waterfall at the mouth of Hurricane River and in a few small 1- to 2-foot exposures at the head of the beach. Better exposures

FIGURE A.16. The Au Sable Light Station at Au Sable Point.

begin about 1/4 to 1/2 mile down the coast to the east and continue uninterrupted along the beach to Au Sable Point and slightly beyond. Those wishing to see the Au Sable Point Lighthouse and maximize their time should consider taking the faster designated trail to Au Sable Point through the woods just inland from the beach (approximately 1.5 miles one way), and then walk the beach back to the parking lot.

Large portions of the beach between Au Sable Point and Hurricane River are developed upon wave-cut benches of the Jacobsville sandstone. Along headlands, these benches are littered with Precambrian crystalline erratics and boulder lags washed from the adjacent drift (Fig. 5.15). At Au Sable Point, the bench extends well out into Lake Superior and is responsible for many of the shipwrecks that litter the shoreline here (Fig. A.17). Excellent exposures of trough cross-bedding (Fig. 2.4) and channel fills (Fig. 2.2) are found throughout the exposure, either at the shoreline or in the low cliffs at the head of the beach. The red colorization has been leached along many of the major joint sets in the Jacobsville Formation, so that the fracture traces are often marked by linear white bands (Fig. 2.3). Rounded splotches of white color, sometimes nearly perfectly spherical, are also common in the Jacobsville, and were formed by localized reducing environments associated with organic or mineral impurities in the sandstone.

FIGURE A.17. Shipwreck along the beach between Hurricane River Campground and Au Sable Point. The Jacobsville sandstone extends far out into the lake here as a shallow bank, creating dangerous conditions for mariners.

Return to car. Retrace road back to H-58.

69.6 miles   Junction H-58. Turn left.

70.1 miles   Road climbs back onto the 690- to 700-foot outwash terrace (216 meters on diagram), designated as the highest Beaver Basin surface (Fig. 4.8d).

70.5 miles   Road climbs up to the 735-foot surface, a small terrace remnant identified as the 222- to 225-meter terrace on Fig. 4.8d.

71.8 miles   Road climbs the impressive Log Slide ice-marginal position, an eastward extension of the Lower Kingston position, onto a small hummocky head of outwash (Fig. 4.8f). This marked the ice position approximately 11,000 years ago during final deglaciation. Road descends this outwash (in some places distinctly kettled) for approximately 2 miles, nearly all the way to the Log Slide turnoff.

GRAND SABLE LAKE 7.5 MINUTE QUADRANGLE MAP

74.1 miles   Junction Log Slide Road. Turn left onto Log Slide Road (paved). Road climbs back up the outwash apron associated with the Log Slide ice-marginal position.

74.8 miles   Log Slide parking area. Park and take short trail to overlook.

Log Slide and Grand Sable Banks and Dunes
This overlook is the best place to view the Grand Sable Banks and Dunes to the east and is considered by many to be one of the most dramatic vistas in Michigan (Fig 1.4). Here the Grand Sable Dunes are perched nearly 400 feet above Lake Superior atop an extensive kame terrace that has been truncated by the modern lake. Only the uppermost portion of the bluff consists of dune sand, with the remainder consisting of outwash sand and gravel and related ice-marginal deposits (valuable information for those wishing to cartwheel down the slope to the lake).

A significant paleosol, the Sable Creek soil (Figs. A.18, 5.2) was developed upon the kame terrace deposits and then buried by the dunes. This soil is well displayed in the bluff beneath and adjacent to the platform upon which you are standing. To reach it, walk just a few hundred yards back toward the parking lot. A path to the left up a small dune leads out to the bluff face, from where the trail descends steeply to the beach. Do not descend to the beach. Instead, just down from the bluff crest, several paleosols can be identified and form weak ledges in the sand to either side of the trail. The best developed of these, up to 12 inches thick in places and located about 20 feet below the dune crest, is the Sable Creek soil.

FIGURE A.18. The Sable Creek soil (arrow), exposed along the top of Grand Sable Banks, about 1/2 mile east of the Log Slide. Tape measure is extended 1 meter (3.3 feet).

As discussed in chapter 5, the dunes are maintained by erosion at the base of the bluff face during high lake levels. Remobilized sand is then blown up the bluff face and deposited in the lee of the crest. During low lake levels, the bluff is stabilized by vegetation and lag gravels until the next high lake phase. Anderton and Loope (1995) concluded that the present dunes were initiated by the rising waters of Lake Nipissing approximately 5,000 to 6,000 years ago. Several episodes of active dune formation followed by periods of stabilization have occurred over the ensuing millennia. The significance of these processes on habitat change within the dunes is described by Loope and McEachern (1998).

The Log Slide gets its name from a large wooden flume that stood at this site during the lumbering period. Logs cut inland were hauled to this site and transported down the bluff to Lake Superior. The logs were then collected into large rafts and transported to sawmills in Grand Marais.

Return to car. Retrace road to H-58.

75.65 miles   Junction of H-58. Turn left.

75.85 miles   Road enters collapsed outwash and begins complicated climbing and descending of various kame terraces all the way to Grand Sable Lake.

76.3 miles   Junction Rhody Creek Truck Trail (on right, Fig. 1.2b). Stay on paved H-58 straight ahead.

78.9 miles   Descend hill from kame terrace down to Sable Lake channelway.

79.1 miles   Turn right into Grand Sable Lake scenic overlook. Follow pavement to small parking area. Park and view lake.

Grand Sable Lake
Grand Sable Lake and its interesting history are discussed in detail in chapter 5. The lake is visible from this vantage point, but the dunes that form its northwestern edge are hidden from view, as is its natural outlet to Lake Superior, Sable Creek (see Sable Falls stop below). During high water levels of Lake Superior, the advancing dune field blocks Sable Creek and Grand Sable Lake expands. When Superior lake levels fall, the dunes stabilize and Sable Creek is able to adjust its channel to drain the lake to a lower level (Loope and others, 2004). An intact stump was found on the lake bottom just to the east of this location, representing forested conditions approximately 3,000 years ago when the lake was low.

Return to car and drive to H-58 junction.

79.3 miles   Junction of H-58. Turn right.

79.7 miles   Grand Sable Lake to the right, Grand Sable Dunes to the left.

80.0 miles   Cross Sable Creek.

80.4 miles   Grand Sable Visitor Center on left. Good view of the 750- to 760-foot (229- to 232-meter) terrace on right side of road (Fig. 3.11).

80.65 miles   Ascend to the 790- to 800-foot (241- to 244-meter) terrace.

80.9 miles   Junction with William Hill Road. Turn left, following the pavement and signs for Grand Marais. Continue on the 790- to 800-foot terrace (Fig. 3.11). Excellent view of terrace surface to the right.

81.4 miles   Descend onto the 750- to 760-foot (229- to 232-meter) terrace (Fig. 3.11). Note the low bedrock outcrop on the right side of the road just before descent.

81.6 miles   Descend onto the 720-foot (219-meter) terrace just before the Sable Falls turnoff.

81.65 miles   Turn left into the Sable Falls parking lot. Park car.

Sable Falls Area

This parking area is the best place to view the stair-step series of kame terraces that developed between the retreating ice margin and the bedrock highlands to the south (Figs. 3.11, 3.12). The parking lot sits on the 720-foot (219-meter) terrace, and the hill south of the parking lot is the riser to the 750-foot (229-meter) terrace. Drainage was from west to east.

Visitors have several trails to choose from at this site. The Dunes or "Ghost Forest" trail leads to several large coastal blowout dunes that previously exposed a "Ghost Forest"—trees from an ancient forest that were buried and then reexposed by the sand. As of 2011 the Ghost Forest was no longer visible, but it may reemerge in the future (Fig. A.19). Hikers can follow the trail across the open dunes to the bluff overlooking Lake Superior. A well-developed paleosol (Sable Creek soil?) is exposed near the top of the bluff.

Another trail leads from the parking lot to Sable Falls and then continues another 0.2 miles to the Lake Superior Beach. This trail is relatively short and worth the walk, although it contains a number

FIGURE A.19. The Ghost Forest as seen in earlier days. (Courtesy of National Park Service)

of stairways. Several fine overlooks allow for excellent views of the 35-foot falls (Fig. A.20), which are formed by Sable Creek as it tumbles toward Lake Superior. According to Haddox (1982), the Jacobsville Formation is exposed at the base of the falls and along the channel bottom downstream. The Basal Conglomerate Member is exposed just above the Jacobsville. The upper 6 feet of the falls are developed upon the Miners Castle Member, with the Chapel Rock Member in between. The contact between the two is covered by vegetation and debris. Thus, nearly the entire stratigraphic

FIGURE A.20. Sable Falls. Nearly the entire stratigraphic section of the park is displayed in "compressed" form at the falls (Haddox, 1982), although access is limited.

section is displayed in "compressed" form at the falls (Haddox 1982). Access to the rocks, however, is restricted by railed walkways.

Red till attributed to the Marquette advance is exposed in the lower part of the Sable Creek valley (Fig. 4.2). This till was likely deposited along the base of the ice at about the time that the heads of outwash of the Kingston Plains (Munising moraine) were being built. To examine the till, begin at the lowermost observation deck for Au Sable Falls. Head lakeward (away from the car) on the wooden walkway approximately 45 paces, climbing several single steps along the way. At the sixth step, look down and left to a meander bend of Sable Creek. The red till is exposed in the cut bank (the steep eastern slope) of the meander.

FIGURE A.21. The Carter organic site, located along the beach east of the mouth of Sable Creek, exposed during high lake levels in the 1990s. Buried dark organic material is exposed at the base of the bluff in the photo, and may represent a paleosurface more than six thousand years old. (Courtesy of Walt Loope, U.S.G.S.)

A significant Holocene organic site, the Carter site, is exposed in the bluffs approximately 3/4 mile east of the mouth of Sable Creek along the beach. This site can be accessed by following the Falls trail out to the shore and then walking eastward along the beach, or by walking westward from the Tourist Park in Grand Marais (see description below). At the time of this writing (2011), the site is no longer visible and has been buried by the slumping bluff. The site is often reexposed during periods of high lake level, however, and distinct layers of dark organic material fill what appears to be a buried pre-Nipissing valley (Fig. A.21). The oldest radiocarbon date on logs recovered from the site is 5,900 years B.P. or about 6,700 calendar years (Loope, personal communication, 2008). A beaver-chewed log (!) from this site (Fig. A.22) was radiocarbon dated at 4,190 ± 40 years B.P. (Loope, personal communication, 2010). A few paleosols with flattened logs are exposed about midway up the bluff and often form weak ledges of darkened sand that can be picked out by a sharp eye from the beach.

Several former channels of Sable Creek described by Loope and colleagues (2004) are located in the woods to the east of the Au Sable Falls trail (for those with limited time, a much more

FIGURE A.22. Beaver-chewed twig, approximately 4,800 calendar years old, collected by Walt Loope from the Carter site near Grand Marais. Scale is in centimeters (2.5 centimeters equals 1 inch). (Courtesy of Walt Loope, U.S.G.S.)

accessible site for viewing these channels is available at Donahey Woods in Grand Marais—see below). They represent times when Sable Creek was displaced eastward by the advancing dunes. To access the former channels here, take the trail from the parking lot toward Au Sable Falls. Just before the stairway leading down to the falls, a cross-country skiing trail (designated "E") leads off to the right. Take this trail to the right. A few hundred feet later, the trail splits. Take the right fork. It parallels the parking lot (in the trees off to the right) and after about 1/2 mile begins to bear left toward the north away from the parking lot. As it nears an old fence line to the right—with a straight line of trees—the path turns right and crosses a wooden bridge over a ditch. Follow the path across the bridge and then walk north (left) along the east side of the ditch to the edge of a dry valley. Descend into the valley and follow it to the right for as long as desired. It eventually is truncated by the modern shoreline at its downstream end. Notice that the channel approximates the size and shape of modern Sable Creek.

Return to car. Drive back to the H-58 turnoff.

82.1 miles   H-58 junction. Turn left. A 720-foot (219-meter) terrace is to the left.

82.3 miles   Road descends to the 680-foot (207-meter) terrace. Terrace visible on both sides of road.

GRAND MARAIS 7.5 MINUTE QUADRANGLE MAP

82.7 miles   Road crosses First Creek, an important player in Holocene stream history (Fig. 3.11). Road rises and turns to the right, then descends into village of Grand Marais. Be ready to turn immediately!

83.0 miles   Junction with Chisholm Avenue. Turn left. Continue straight ahead one block to corner of Brazel Street. Proceed straight ahead to a small parking area.

Park car. Take the path west from the parking lot to a descending stairway.

Donahey Woods Site

Burt Township acquired this land from the Donahey family with the assistance of the Nature Conservancy. The stairway descends to a railed deck built in an abandoned late Holocene meander of First Creek. A second stairway then leads down to the Lake Superior beach, along a stretch particularly popular with beachcombers and agate hunters. The Carter organic site can be reached by walking the beach westward from here approximately 1/3 mile.

Descend the first stairway to the observation deck. The deck sits on a well-developed incised meander that trends from left to right as a conspicuous, dry (usually), flat-bottomed valley (Fig. A.23). Its downstream end has been truncated by the Lake Superior bluff about 10 to 15 feet above the modern beach, suggesting that the meander may be related to the Algoma (or possibly a post Algoma) phase of Lake Superior. A short but worthwhile walk up the channel to the left brings you to a steep declivity where the abandoned meander hangs dramatically above the valley of First Creek.

FIGURE A.23. An impressive dry paleovalley of First Creek located in Donahey Woods near Grand Marais. Note person for scale. (Photo by Gretchen Kuhn)

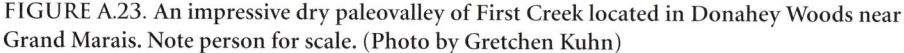

After returning to the observation deck, proceed down the second stairway to the Lake Superior Beach for beachcombing or a quick nap.

Return to car.

83.1 miles   From parking lot, turn left onto Brazel Street. The Tourist Park Campground is on your left.

83.45 miles   Stop at Gitche Gumee Agate and History Museum on right side of the road. The gift shop contains an excellent display of agates (both local and exotic) as well as agate hunting supplies. The museum features exhibits on local geology. Well worth the stop and a great source for local geologic information.

83.5 miles   Junction with Lake Avenue. Turn left and descend small hill (Algoma beach scarp).

83.6 miles   Turn right into the Burt Township Marina.

From this site, abandoned shorelines related to the fall from Lake Nipissing and the Algoma shoreline are visible. The bluff along the base of the West Bay here is developed upon the combined effects of the Algoma and another, slightly older and higher shoreline bluff (Fig. 3.15). These shorelines enter the village from the east as separate scarps (Fig. 3.16), merging just east of downtown. As discussed in chapter 3, the Grand Marais business district sits on old Lake Nipissing bottom (Fig. 3.14). The hill at the south end of town marks the former Lake Nipissing storm beach (Fig. 3.11).

Return to car. From the parking lot, turn right onto the main road. Follow to the circle at end of road.

84.3 miles   Turn right from the circle into the parking area at the Grand Marais breakwall. Park and walk toward the pier.

Grand Marais

The tiny village of Grand Marais grew and withered with the fortunes of the logging industry and today serves as an exceedingly picturesque tourist destination and eastern gateway for Pictured Rocks National Lakeshore. As the only harbor of refuge between Sault Ste. Marie and Marquette, Grand Marais Harbor has long been a resting place for Native Americans, voyageurs, and Great Lakes mariners. Before the construction of the breakwall, Grand Marais harbor was separated from the open lake by a low sandy baymouth bar that extended from the Coast Guard station eastward to Lonesome Point (the low promontory in the distance marking the eastern end of the bay) (Regis and Anderton, 1999). A small channel, easily navigable by small boat, connected the bay to the open waters of Lake Superior. This channel was prone to sandbar development, however, and interfered with commercial shipping. Beginning in 1883, a series of harbor protection structures were built to stop the formation of sandbars,

FIGURE A.24. Oblique aerial view of Grand Marais Harbor. View is southwestward. (Courtesy of National Park Service)

culminating in the present breakwater structure. Unfortunately, these structures blocked the natural currents that provided sand to the eastern parts of the original barrier. As a consequence, the bar protecting this side of the harbor has been eroded away, exposing nearly the entire eastern side of Grand Marais Harbor to the open lake (Fig. A.24). Because sand is carried from west to east along the shoreline from source areas in the Grand Sable Banks, the beach on the western side of the breakwall has continued to build, growing at a rate of over 1.3 million cubic feet (38,000 cubic meters) per year (Regis and Anderton, 1999). If this expansion continues unchecked, the accumulating sand will eventually cut off wave contact with the Grand Sable Banks, effectively stabilizing the dunes (Marsh, 1990). Meanwhile, some sections of the bay have been shoaling with sand carried around the present breakwater, alarming local residents. As this book went to press, consultants have offered several remediation strategies to address these problems, and in June 2011, after an intensive lobbying effort by local citizenry, the state of Michigan announced plans to construct a new breakwall to better protect the harbor.

End of Tour

REFERENCES CITED

Ahlgren, C. E., and Ahlgren, I. F., 1983, The human impact on northern forest ecosystems, in Flader, S. F., ed., The Great Lakes forest: A social and environmental history: St. Paul: University of Minnesota Press, p. 33–51.

Aitken, M. J., 1998, An introduction to optical dating: Oxford: Oxford University Press, p. 267.

Anderton, J. B., 2002, The Miners area, Pictured Rocks National Lakeshore, in Loope, W. L., and Anderton, J. B. eds., Deglaciation and Holocene landscape evolution in eastern upper Michigan, a field guide for the 48th Midwest Friends of the Pleistocene field conference, May 31–June 2, Grand Marais, Michigan, p. 43–46.

Anderton, J., Legg, R., and Regis, R., 2009, Geoarchaeological approaches to site location modeling and archaeological survey in the Pictured Rocks National Lakeshore, Michigan: Wisconsin Archeologist, v. 90, nos. 1 and 2, p. 19–30.

Anderton, J. B., and Loope, W. L., 1995, Buried soils in a perched dunefield as indicators of late Holocene lake-level change in the Lake Superior Basin: Quaternary Research, v. 44, p. 190–199.

Baedke, S. J., and Thompson, T. A., 2000, A 4,700-year record of lake level and isostasy for Lake Michigan: Journal of Great Lakes Research, v. 26, no. 4, p. 416–426.

Barrett, L. R., 1995, A stump prairie landscape in northern Michigan: Soils, forest, vegetation, logging, and fire: East Lansing, Michigan State University, Ph.D. dissertation.

———1997, Podzolization under forest and stump prairie vegetation in northern Michigan: Geoderma, v. 78, p. 37–58.

Barrett, L. R., and Schaetzl, R. J., 1998, Regressive pedogenesis following a century of deforestation: Evidence for depodzolization: Soil Science, v. 163, p. 482–497.

Bates, R. L., and Jackson, J. A., 1987, Glossary of geology (3d ed.): Alexandria, Va.: American Geological Institute.

Benchley, E. D., Marcucci, C. Yuen, and Griffin, K., 1988, Final report of the archaeological investigations and data recovery at the Trout Point I site, Alger County, Michigan: Report of Investigations 89, Archaeological Research Laboratory, University of Wisconsin–Milwaukee.

Bergquist, S. G., 1936a, The Grand Sable Dunes on Lake Superior, Alger County, Michigan: Papers of the Michigan Academy of Science, Arts, and Letters, v. 21, p. 429–438.

———1936b, The Pleistocene history of the Tahquamenon and Manistique Drainage region of the Northern Peninsula of Michigan: Michigan Geological Survey Publication 40, part 1, p. 7–148.

Blewett, W. L., 1984, Ice stagnation landforms in Eastern Upper Michigan: A reinterpretation of the Munising moraine. Macomb, Illinois, Western Illinois University, unpublished master's thesis.

———1990, The glacial geomorphology of the Port Huron Complex in northwestern southern Michigan: East Lansing, Michigan State University, Ph.D. dissertation.

———1994, Late Wisconsin history of Pictured Rocks National Lakeshore and vicinity: Pictured Rocks Resource Report #94-01: National Park Service, Washington, D.C.: U.S. Government Printing Office, available at Pictured Rocks National Lakeshore Headquarters, Munising, Michigan.

———2002a, Glacial geomorphology of Pictured Rocks National Lakeshore and vicinity, in Loope, W. L., and Anderton, J. B., eds., Deglaciation and Holocene landscape evolution in eastern upper Michigan, field guide for the 48th Midwest Friends of the Pleistocene field conference, May 31–June 2, Grand Marais, Michigan, p. 11–21.

———2002b, Late Wisconsin history of Eastern Upper Michigan: A review, in Loope, W. L., and Anderton, J. B., eds., Deglaciation and Holocene landscape evolution in eastern upper Michigan, field guide for the 48th Midwest Friends of the Pleistocene field conference, May 31–June 2, Grand Marais, Michigan, p. 1–10.

———2009, Understanding ancient shorelines in the national parklands of the Great Lakes. Pictured Rocks Resource Report #09-01, National Park Service, available at Pictured Rocks National Lakeshore Headquarters, Munising, Michigan.

Blewett, W. L., Lusch, D. P., and Schaetzl, R. J., 2009, The physical landscape: A glacial legacy, in Schaetzl, R. J., Darden, J., and Brandt, D., eds., Michigan geography and geology: New York: Pearson Custom Publishing, p. 249–273.

Blewett, W. L., and Rieck, R. L., 1987, Reinterpretation of a portion of the Munising moraine in northern Michigan: Geological Society of America Bulletin, v. 98, p. 169–175.

Blewett, W. L., and Winters, H. A., 1995, The importance of glaciofluvial features within Michigan's Port Huron moraine: Annals of the Association of American Geographers, v. 85, no. 2, p. 306–319.

Booth, R. K., Jackson, S. T., and Thompson, T. A., 2002, Paleoecology of a northern Michigan lake and the relationship among climate, vegetation, and Great Lakes water levels: Quaternary Research, v. 57, p. 120–130.

Brzys, K., 2004, Understanding and finding agates: Hancock, Mich.: Book Concern Printers.

Castaneda, M., Fisher, T. G., Jol, H. M., Loope, W., Campbell, M., and Goble, R., 2009, Origin and age of Sand Point, Pictured Rocks National Lakeshore, Michigan, USA: International Association of Great Lakes Research, Annual Meeting, Toledo, Ohio, May 18–22.

Catacosinos, P. A., and Daniels, P. A., Jr., 1991, Stratigraphy of Middle Proterozoic to Middle Ordovician formations in the Michigan Basin, in Catacosinos, P. A., and Daniels, P. A., Jr., eds., Early sedimentary evolution of the Michigan Basin: Geological Society of America Special Paper 256, p. 53–71.

Clague, J., 1975, Sedimentology and paleohydrology of Late Wisconsin outwash, Rocky Mountain Trench, southeastern British Columbia, in Jopling, A., and McDonald, B., eds., Glaciofluvial and glaciolacustrine sedimentation: Society of Economic Paleontologists and Mineralogists Special Publication 23, p. 223–237.

Clark, C., 1993, Archaeological survey and testing at Pictured Rocks National Lakeshore, Alger County, Michigan 1991: Lincoln, Neb.: Midwest Archaeological Center Technical Report No. 23, U.S. Department of the Interior, National Park Service.

Clayton, L., 1983, Chronology of Lake Agassiz drainage to Lake Superior, in Teller, J. T., and Clayton, L., eds., Glacial Lake Agassiz: Geological Association of Canada Special Paper 26, p. 291–307.

Clayton, L., and Moran, S., 1974, A glacial process form model, in Coates, D. R., ed., Glacial geomorphology: Binghamton, N.Y.: State University of New York at Binghamton Publications in Geomorphology, p. 89–119.

Cleland, C. E., 1992, Rites of conquest: The history and culture of Michigan's Native Americans: Ann Arbor: University of Michigan Press.

Craft, E., Jol, H., Fisher, T. G., Loope, W., Campbell, M., and Castaneda, M., 2009, Ground Penetrating Radar image of a cuspate foreland, Sand Point, Michigan: Association of American Geographers Annual Meeting, Las Vegas, Nevada, March 22–27.

Derouin, S. A., and Regis, R. S., 2005, Using gravity methods to determine the existence and extension of tunnel channels across the Upper Peninsula of Michigan: Geological Society of America annual meeting, Abstracts with Programs, October 2005, v. 37, no. 7, p. 522.

Dorr, J. A., Jr., and Eschman, D. F., 1970, Geology of Michigan: Ann Arbor: University of Michigan Press.

Drexler, C. W., 1975, Report on the geology of the Pictured Rocks National Lakeshore, on file at the headquarters of Pictured Rocks National Lakeshore, Munising, Michigan.

———1981, Outlet channels for the post-Duluth lakes of the Upper Peninsula of Michigan: Ann Arbor, University of Michigan, Ph.D. dissertation.

Drexler, C. W., Farrand, W. R., and Hughes, J. D., 1983, Correlation of glacial lakes in the Superior basin with eastward discharge events from Lake Agassiz, in Teller, J. T., and Clayton, L., eds., Glacial Lake Agassiz: Geological Association of Canada Special Paper 26, p. 309–329.

Driscoll, E. G., Jr., 1956, An environmental and heavy mineral study of the "Eastern Sandstones" between Marquette and Grand Marais, Michigan: Madison, University of Wisconsin, unpublished master's thesis.

Dunham, S., and Anderton, J. B., 1999, Late Archaic radiocarbon dates from the Popper site (FS 09-10-03-825/20 AR 350): A multicomponent site on Grand Island, Michigan: Michigan Archaeologist, v. 45, no. 1, p. 1–22.

Dury, G. H., 1965, Theoretical implications of underfit streams: U.S. Geological Survey Professional Paper 452-C.

Fairbanks, R. G., Mortlock, R. A., Chiu, T. C., Cao, L., Kaplan, A., Guilderson, T. P., Fairbanks, T. W., and Bloom, A. L., 2005, Marine radiocarbon calibration curve spanning 10,000 to 50,000 years B.P. based on paired 230 Th/234U/238U and 14C dates on pristine corals: Quaternary Science Reviews, v. 24, p. 1781–1796.

Farrand, W. R., and Bell, D. L., 1982, Quaternary geology of Michigan: Michigan Department of Natural Resources, map, 1:500,000 scale.

Farrand, W. R., and Drexler, C. W., 1985, Late Wisconsinan and Holocene history of the Lake Superior basin, in Karrow, P. F., and Calkin, P. E., eds., Quaternary evolution of the Great Lakes: Geological Association of Canada Special Paper 30, p. 18–32.

Farrell, J. P., and Hughes, J. D., 1984, Wave erosion and mass wasting: Pictured Rocks National Lakeshore, p. 8–26, report available at Pictured Rocks National Lakeshore Headquarters.

————1985, Long term implications, from a geomorphological standpoint, of maintaining H-58 in its present location at Grand Sable Lake: National Park Service Contract Report 60000-4-0839: Marquette, Mich.: Department of Geography, Earth Science, Conservation, and Planning, Northern Michigan University, available at Pictured Rocks National Lakeshore Headquarters.

Fisher, T. G., 2008, Determining the origin and dynamics of coastal processes of Sand Point at Pictured Rocks National Lakeshore: Final Technical Report, Award #07HQAG0150, 13 p., available at Pictured Rocks National Lakeshore Headquarters.

Fisher, T. G., and Whitman, R. L., 1999, Deglacial and lake level history recorded in cores, Beaver Lake, Upper Peninsula, Michigan: Journal of Great Lakes Research, v. 25, no. 2, p. 263–274.

Flint, R. F., 1971, Glacial and Quaternary geology: New York: John Wiley and Sons.

Foster, J. W., and Whitney, J. D., 1851, Report on the geology of the Lake Superior land district; part 2, the iron region, together with the general geology: U.S. 32nd Congress, Special Session, Senate Executive Documents, No. 4.

Fraser, G. S., Larsen, C. E., and Hester, N. C., 1990, Climatic control of lake levels in the Lake Michigan and Lake Huron basins, in Schneider, A. F., and Fraser, G. S., eds., Late Quaternary history of the Lake Michigan basin: GSA Special Paper 251, p. 75–90.

Frederick, D. J., Rakestraw, L., Christopher, R. E., and Van Dyke, R. A., 1976, Original forest vegetation of the Pictured Rocks National Lakeshore and a comparison with present conditions: Michigan Academician, v. 9, p. 433–443.

Futyma, R. P., 1981, The northern limits of glacial Lake Algonquin in upper Michigan: Quaternary Research, v. 15, p. 291–310.

Gauthier, K. J., and Mueller, B. E., 2007, Lake Superior rock picker's guide: Ann Arbor: University of Michigan Press, and Traverse City: Petoskey Publishing Company.

Grabau, A. W., 1906, Types of sedimentary overlap: Geological Society of America Bulletin, v. 17, p. 567–636.

Guldzenopf, E. C., 1967, Conodonts from the Prairie du Chien of Northern Michigan—preliminary report, in Ostrom, M. E., and Slaughter, A. E., eds., Correlation problems of the Cambrian and Ordovician outcrop areas, Northern Peninsula of Michigan: Michigan Basin Geological Society Annual Field Excursion Guidebook, p. 58–64.

Haddox, C. A., 1982, Sedimentology of the Munising Formation: Madison, University of Wisconsin, unpublished master's thesis.

Haddox, C. A., and Dott, R. H., 1990, Cambrian shoreline deposits in northern Michigan: Journal of Sedimentary Petrology, v. 60, p. 697–716.

Hamblin, W. K., 1958, Cambrian sandstones of northern Michigan: Michigan Department of Conservation, Geological Survey Division, Publication 51.

Hansel, A. K., Mickelson, D. M., Schneider, A. F., and Larsen, C. E., 1985, Late Wisconsinan and Holocene history of the Lake Michigan basin, in Karrow, P. F., and Calkin, P. E., eds., Quaternary evolution of the Great Lakes: Geological Association of Canada Special Paper 30, p. 39–53.

Holm, D. K., Schneider, D. A., Rose, S., Mancusco, C., McKenzie, M., Foland, K. A., and Hodges, K. V., 2007, Proterozoic metamorphism and cooling in the southern Lake Superior region, North America, and its bearing on crustal evolution: Precambrian Research, v. 157, p. 71–79.

Hough, J. L., 1958, Geology of the Great Lakes: Urbana: University of Illinois Press.

Houghton, D., 1841, Fourth annual report of the state geologist: Michigan House of Representatives Document No. 27, p. 3–89.

Hughes, J. D., 1968, Unpublished materials for the geology section of the Pictured Rocks National Lakeshore Master Plan, available at the Pictured Rocks National Lakeshore Headquarters.

———1971, Post-Duluth stage outlet from the Lake Superior basin: Michigan Academician, v. 3, p. 71–77.

———1978, Marquette buried forest 9,850 years old: Abstract for the American Association for the Advancement of Science annual meeting, February 12–17.

———1989, When Green Bay was a valley: The Au Train–Whitefish–Green Bay spillway, in Palmquist, J. C., ed., Wisconsin's Door Peninsula: A natural history: Appleton, Wis.: Perin Press.

Johnston, J. W., Baedke, S. J., and Thompson, T. A., 2002, Beach-ridge architecture, lake level change, and shoreline behavior in the Au Train embayment, Au Train, Michigan, in Loope, W. L., and Anderton, J. B., eds., Deglaciation and Holocene landscape evolution in eastern upper Michigan, field guide for the 48th Midwest Friends of the Pleistocene field conference, May 31–June 2, Grand Marais, Michigan, p. 49–55.

Johnston, J. W., Thompson, T. A., and Baedke, S. J., 2000, Preliminary report of late Holocene lake-level variation in southern Lake Superior: part 1: Indiana Geological Survey Open-File Study, p. 99–118.

Johnston, J. W., Thompson, T. A., Wilcox, D. A., and Baedke, S. J., 2007, Geomorphic and sedimentologic evidence for the separation of Lake Superior from Lake Michigan and Huron: Journal of Paleolimnology, v. 37, p. 349–364.

Kalliokoski, J., 1982, Jacobsville sandstone, in Wold, R. J., and Hinze, W. J., eds., Geology and tectonics of the Lake Superior Basin: Geological Society of America Memoir 156, p. 147–155.

Karamanski, T. J., 1989, Deep woods frontier: A history of logging in Northern Michigan: Detroit: Wayne State University Press.

————1995, The Pictured Rocks: An administrative history of Pictured Rocks National Lakeshore: Omaha: Midwest Regional Office, National Park Service, U.S. Department of the Interior.

Karrow, P. F., Anderson, T. W., Clarke, A. H., DeLorme, L. D., and Sreenivasa, M. R., 1975, Stratigraphy, paleontology, and age of Lake Algonquin sediments in southwestern Ontario, Canada: Quaternary Research, v. 5, p. 49–87.

Kincare, K., and Larson, G. J., 2009, Evolution of the Great Lakes, in Schaetzl, R. J., Darden, J., and Brandt, D., eds., Michigan geography and geology: New York: Pearson Custom Publishing, p. 174–190.

Koteff, C., and Pessl, F., 1981, Systematic ice retreat in New England: U.S. Geological Survey Professional Paper 1179.

Lane, A. C., and Seaman, A. E., 1907, Notes on the geological section of Michigan, part 1: The pre-Ordovician: Journal of Geology, v. 15, p. 680–695.

Larsen, C. E., 1985, Lake level, uplift, and outlet incision, the Nipissing and Algoma Great Lakes, in Karrow, P. F., and Calkin, P. E., eds., Quaternary evolution of the Great Lakes: Geological Association of Canada Special Paper 30, p. 63–77.

————1987, Geological history of glacial Lake Algonquin and the Upper Great Lakes: U.S. Geological Survey Bulletin 1801.

Larson, G. J., and Kincare, K., 2009, Late Quaternary history of the eastern mid-continent region, USA, in Schaetzl, R. J., Darden, J., and Brandt, D., eds., Michigan geography and geology: New York: Pearson Custom Publishing, p. 69–90.

Legg, R., and Anderton, J., 2010, Using paleoshoreline and site location modeling in the Northern Great Lakes: Geoarchaeological approaches to prehistoric archaeological survey in the Pictured Rocks National Lakeshore: Geoarchaeology: An International Journal, v. 25, no. 6, p. 772–783.

Leverett, F., 1911, Map of the surface formations of the Northern Peninsula of Michigan: Michigan Department of Conservation, Geological Survey Division, 1:380,160 scale.

————1917, Surface geology and agricultural conditions of Michigan: Michigan Department of Conservation, Geological Survey Division, Publication 25, Geology series 21.

————1929, Moraines and shorelines of the Lake Superior region: U.S. Geological Survey Professional Paper 154-A.

Levin, H. L., 1999, The Earth through time (6th ed.): Fort Worth, Tex.: Saunders College Publishing.

Loope, W. L., 1993, Evidence of physical and biological change within the Beaver Lake watershed attributable to a turn-of-the-century logging dam: Pictured Rocks Resource Report #93-02, National Park Service, Washington, D.C.: U.S. Government Printing Office.

Loope, W. L., Fisher, T. G., Jol, H. M., Goble, R. J., Anderton, J. B., and Blewett, W. L., 2004, A Holocene history of dune-mediated landscape change along the southeastern shore of Lake Superior: Geomorphology, v. 61, p. 301–322.

Loope, W. L., and Holman, M. P., 1991, An assessment of stream bed and stream bank characteristics within Pictured Rocks National Lakeshore: Pictured Rocks Resource Report #91-01, National Park Service, Washington, D.C.: U.S. Government Printing Office.

Loope, W. L., and McEachern, A. K., 1998, Habitat change in a perched dune system, in Mac, M. J., Opler, P. A., Puckett Haecker, C. E., and Doran, P. D., eds., Status and trends of the nation's biological resources, vol. 1: U.S. Department of the Interior, U.S. Geological Survey, Reston, Virginia, p. 227–230.

Lovis, W. A., 2009, Between the glaciers and Europeans: People from 12,000 to 400 years ago, in Schaetzl, R. J., Darden, J., and Brandt, D., eds., Michigan geography and geology: New York: Pearson Custom Publishing, p. 389–401.

Lowell, T. V., Larson, G. L., Hughes, J. D., and Denton, G. H., 1999, Age verification of the Lake Gribben forest bed and the Younger Dryas Advance of the Laurentide Ice Sheet: Canadian Journal of Earth Science, v. 36, p. 383–393.

Lynch, D. R., and Lynch, B., 2010, Michigan rocks and minerals: A field guide to the Great Lake State: Cambridge, Minn.: Adventure Publications, Inc.

Lytle, D. E., 2005, Palaeoecological evidence of state shifts between forest and barrens on a Michigan sand plain, USA: The Holocene, v. 15, p. 821–836.

Macpherson, H. G., 1989, Agates: London: British Museum of Natural History.

Maizels, J., 1983, Channel changes, paleohydrology, and deglaciation: Evidence for some late glacial sandur deposits of northeast Scotland: Quaternary Studies in Poland, no. 4, p. 171–187.

Marsh, W. M., 1990, Nourishment of perched sand dunes and the issue of erosion control in the Great Lakes: Environmental Geology and Water Sciences, v. 16, no. 2, p. 155–164.

Marsh, W. M., and Marsh, B. D., 1987, Wind erosion and sand dune formation on high Lake Superior bluffs: Geografiska Annaler, v. 69A, no. 3–4, p. 379–391.

Martin, H. M., 1957, Map of the surface formations of the Northern Peninsula of Michigan: Michigan Department of Conservation, Geological Survey Division, 1:500,000 scale.

McKee, R., 1988, Tombstones of a lost forest: Audubon, v. 90, no. 2, p. 62–72.

Mickelson, D. M., Clayton, L., Fullerton, D. S., and Borns, H. W., Jr., 1983, The late Wisconsin glacial record of the Laurentide ice sheet in the United States, in Porter, S. C., ed., The late Pleistocene: Minneapolis: University of Minnesota Press, p. 3–37.

Miller, J. F., Ethington, R. L., and Rosé, R., 2006, Stratigraphic implications of Lower Ordovician conodonts from the Munising and Au Train Formations of Pictured Rocks National Lakeshore, Upper Peninsula of Michigan: Palaios, v. 21, p. 227–237.

Milstein, R. L., 1987, Pictured Rocks National Lakeshore, northern Michigan, in Biggs, D. L., ed., Geological Society of America Centennial Field Guide—North-Central section, p. 277–280.

Mooers, H. D., 1990, A glacial-process model: The role of spatial and temporal variations in glacier thermal regime: Geological Society of America Bulletin, v. 102, p. 243–251.

NICE (Northern Interior Continental Evolution) Working Group: Holm, D. K., Anderson, R., Boerboom, T. J., Cannon, W. F., Chandler, V., Jirsa, M., Miller, J., Schneider, D. A., Schulz, K. J., and Van Schmus, W. R., 2007, Reinterpretation of Paleoproterozoic accretionary boundaries of the north-central United States based on a new aeromagnetic-geologic compilation: Precambrian Research, v. 157, p. 71–79.

Niemi, A. A., 1973, Michigan's glacial gemstones of [the] northeastern Upper Peninsula, self-published.

Oetking, P. F., 1951, The relation of the Lower Paleozoic to the older rocks in the northern peninsula of Michigan: Madison, University of Wisconsin, Ph.D. dissertation.

Ostrom, M. E., and Slaughter, A. E., eds., 1967, Correlation problems of the Cambrian and Ordovician outcrop areas, Northern Peninsula of Michigan: Michigan Basin Geological Society Annual Field Excursion Guidebook.

Pabian, R. K., and Zarins, A., 1994, Banded agates: Origins and inclusions: University of Nebraska Educational Circular 12, Lincoln: University of Nebraska.

Prest, V. K., Donaldson, T. A., and Mooers, H. D., 2000, The Omar story: The role of omars in assessing the glacial history of west central North America: Geographie Physique et Quaternaire, v. 54, p. 257–270.

Regis, R. S., and Anderton, J. B., 1999, Paleozoic and glacial geology from Au Train to Grand Marais, Michigan, in Bornhorst, T. J., ed., Institute on Lake Superior Geology Field Trip Guidebook Proceedings, Field trip no. 4, v. 45, 45th annual meeting, Marquette, Michigan, May 4–8.

Regis, R. S., Jennings-Patterson, C., Wattrus, N., and Rausch, D., 2003, Relationship of deep troughs in the eastern Lake Superior basin and large-scale glaciofluvial landforms in the central Upper Peninsula of Michigan: Geological Society of America North-Central section annual meeting, Abstracts with Programs, February 2003, v. 35, no. 2, p. 55.

Regis, R. S., Loope, W. L., and Goble, R. J., 2004, Buried tunnel valleys across the Upper Peninsula of Michigan: Groundwater conduits for outburst floods from glacial Lake Agassiz?: Geological Society of America annual meeting, Abstracts with Programs, November 2004, v. 36, no. 5, p. 68.

Rieck, R. L., 1991, Map of Alger County bedrock topography, unpublished map.

Robinson, S., 2001, Is this an agate?: An illustrated guide to Lake Superior's beach stones [in] Michigan: Hancock, Mich.: Book Concern Printers.

Rosé, R., 1997, Overview of Cambrian sandstone environments of deposition: Pictured Rocks Resource Report #97-01, National Park Service, Washington, D.C.: U.S. Government Printing Office.

Roy, J. L., and Robertson, W. A., 1978, Paleomagnetism of the Jacobsville Formation and the apparent polar path for the interval 1100–670 million years for North America: Journal of Geophysics Research, v. 83, p. 1289–1304.

Schulz, K. J., and Cannon, W. F., 2007, The Penokean Orogeny in the Lake Superior Region: Precambrian Research, v. 157, no. 1–4, p. 4–25.

Shackelton, N. J., and Hall, M. A., 1984, Oxygen and carbon isotope stratigraphy of Deep Sea Drilling Project Hole 552A: Plio-Pleistocene glacial history, in Roberts, D. G., and others, eds., Initial report of the Deep Sea Drilling Project: Washington, D.C.: U.S. Government Printing Office, v. 81, p. 599–609.

Shaub, B. M., 1979, Genesis of thundereggs, geodes, and agates of igneous origin: Lapidary Journal, v. 32, no. 11, p. 2340–2354.

Sloss, S. L., 1963, Sequences in the cratonic interior of North America: Geological Society of America Bulletin, v. 74, p. 93–111.

Spencer, J. W., 1891, Origin of the basins of the Great Lakes of America: American Geologist, v. 7, p. 86–97.

Teller, J. T., and Thorleifson, L. H., 1983, The Lake Agassiz–Lake Superior connection, in Teller, J. T., and Clayton, L., eds., Glacial Lake Agassiz: Geological Association of Canada Special Paper 26, p. 261–290.

Thompson, T. A., and Baedke, S. J., 1995, Beach ridge development in Lake Michigan—shoreline behavior in response to quasi-periodic lake-level events: Marine Geology, v. 129, p. 163–174.

Thompson, T. A., and Baedke, S. J., 1997, Strand-plain evidence for late Holocene lake-level variations in Lake Michigan: Geological Society of America Bulletin, v. 109, no. 6, p. 666–682.

Thornbury, W. D., 1965, Regional geomorphology of the United States: New York: John Wiley and Sons.

Thwaites, F. T., 1934, Well logs in the northern peninsula of Michigan showing the Cambrian section: Michigan Academy of Science, Arts, and Letters Papers, v. 19, p. 413–426.

Vogel, J., 2000, History of fish and fisheries in the Pictured Rocks National Lakeshore: Pictured Rocks Resource Report #2000-1, National Park Service, Washington, D.C.: U.S. Government Printing Office.

Walker, J. D., and Geissman, J. W., compilers, 2009, Geologic time scale: Geological Society of America: GSA Today, v. 19, no. 4/5, p. 61.

Wang, Y., and Merino, E., 1990, Self-organizational origin of agates: Banding, fiber twisting, composition, and dynamic crystallization model: Geochimica et Cosmochimica Acta, v. 54, no. 6, p. 1627–1638.

Wolter, S. F., 2008, The Lake Superior agate (4th ed.): Minneapolis, self-published.

Young, R., 2004, An analysis of coastal erosion and management issues at Sand Point, Pictured Rocks National Lakeshore: Technical report submitted to Pictured Rocks National Lakeshore, August 23, 2004, available at the Pictured Rocks National Lakeshore Headquarters.

INDEX